Springer Tracts in Advanced Robotics

Volume 156

The Springer Tracts in Advanced Robotics (STAR) publish new developments and advances in the fields of robotics research, rapidly and informally but with a high quality. The intent is to cover all the technical contents, applications, and multidisciplinary aspects of robotics, embedded in the fields of Mechanical Engineering, Computer Science, Electrical Engineering, Mechatronics, Control, and Life Sciences, as well as the methodologies behind them. Within the scope of the series are monographs, lecture notes, selected contributions from specialized conferences and workshops, as well as selected Ph.D. theses.

Special offer: For all clients with a print standing order we offer free access to the electronic volumes of the Series published in the current year.

Indexed by SCOPUS, DBLP, EI Compendex, zbMATH, SCImago.

All books published in the series are submitted for consideration in Web of Science.

Michael Lutter

Inductive Biases in Machine Learning for Robotics and Control

Springer

Michael Lutter
Darmstadt, Hessen, Germany

ISSN 1610-7438 ISSN 1610-742X (electronic)
Springer Tracts in Advanced Robotics
ISBN 978-3-031-37831-7 ISBN 978-3-031-37832-4 (eBook)
https://doi.org/10.1007/978-3-031-37832-4

This Springer imprint is published by the registered company Springer Nature Switzerland AG
The registered company address is: Gewerbestrasse 11, 6330 Cham, Switzerland

Series Editor's Foreword

At the dawn of the century's third decade, robotics is reaching an elevated level of maturity and continues to benefit from the advances and innovations in its enabling technologies. These innovations are all contributing to an unprecedented effort to bringing robots to human environment in hospitals and homes, factories, and schools; in the field for robots fighting fires, making goods and products, picking fruits, and watering the farmland, saving time and lives. Robots today hold the promise of making a considerable impact in a wide range of real-world applications from industrial manufacturing to healthcare, transportation, and exploration of the deep space and sea. Tomorrow, robots will become pervasive and touch upon many aspects of modern life.

The *Springer Tracts in Advanced Robotics (STAR)* is devoted to bringing to the research community the latest advances in the robotics field on the basis of their significance and quality. Through a wide and timely dissemination of critical research developments in robotics, our objective with this series is to promote more exchanges and collaborations among the researchers in the community and contribute to further advancements in this rapidly growing field.

The monograph by Michael Lutter is based on the author's doctoral thesis. The contents are focused on machine learning for robot control and are organized into five chapters. Three different approaches are introduced with the goal of incorporating existing knowledge as inductive biases in machine learning algorithms. Each of these algorithms is evaluated on physical systems and compared to the classical engineering and deep learning baselines.

The work is rich of experiments highlighting the increase in performance of inductive biases compared to black-box deep learning approaches. This volume was the winner of the 2022 Georges Giralt Ph.D. Award for the best doctoral thesis in Europe. A very fine addition to the STAR series!

Naples, Italy
May 2023

Bruno Siciliano
STAR Editor

Preface

A fundamental problem of robotics is how can one program a robot to perform a task with its limited embodiment? Classical robotics solves this problem by carefully engineering interconnected modules. The main disadvantage is that this approach is labor-intensive and becomes close to impossible for unstructured environments and observations. Instead of manual engineering, one can solely use black-box models and data. In this paradigm, interconnected deep networks replace all modules of classical robotics. The network parameters are learned using reinforcement learning or self-supervised losses that predict the future.

In this book, I want to show that these two approaches of classical engineering and black-box deep networks are not mutually exclusive. One can transfer insights from classical robotics to the black-box deep networks and obtain better learning algorithms for robotics and control. To show that incorporating existing knowledge as inductive biases in machine learning algorithms can improve performance, three different algorithms are presented: (1) The Differentiable Newton-Euler Algorithm (DiffNEA) reinterprets the classical system identification of rigid bodies. By leveraging automatic differentiation, virtual parameters, and gradient-based optimization, this approach guarantees physically consistent parameters and applies to a wider class of dynamical systems. (2) Deep Lagrangian Networks (DeLaN) combine deep networks with Lagrangian mechanics to learn dynamics models that conserve energy. Using two networks to represent the potential and kinetic energy enables the computation of a physically plausible dynamics model using the Euler-Lagrange equation. (3) Robust Fitted Value Iteration (rFVI) leverages the control-affine dynamics of mechanical systems to extend value iteration to the adversarial reinforcement learning with continuous actions. The resulting approach enables the computation of the optimal policy that is robust to changes in the dynamics.

Each of these algorithms is evaluated on physical systems and compared to the classical engineering and deep learning baselines. The experiments show that the inductive biases increase performance compared to black-box deep learning approaches. DiffNEA solves Ball-in-Cup on the physical Barrett WAM using offline model-based reinforcement learning with only four minutes of data. The deep network models fail on this task despite using more data. DeLaN obtains a model that can be used for energy control of under-actuated systems. Black-box models cannot be applied as these cannot infer the system energy. rFVI learns robust policies that can swing up the Furuta pendulum and cartpole. The rFVI policy is more robust to changes in the pendulum mass compared to deep reinforcement learning with uniform domain randomization.

In conclusion, this book introduces the combination of prior knowledge and deep learning. The presented algorithms highlight that one can use deep networks in more creative ways than naive input-output mappings for dynamics models and policies. Compared to the deep learning baselines, the proposed approaches can be applied to more problems and improve performance.

Cambridge, Darmstadt
April 2023

Michael Lutter

Acknowledgements

During my Ph.D., I had the pleasure to work with amazing researchers that helped me grow and improve my research. Therefore, I would like to thank the researchers that contributed to my Ph.D. thesis:

- My Ph.D. supervisor Jan Peters, who provided an excellent environment for me to complete my thesis works.
- My external collaborators: Arunkumar Byravan, Debora Clever, Gabe Dulac-Arnold, Animesh Garg, Leonard Hasenclever, Nicolas Heess, Kim Listmann, Shie Mannor, Josh Merel, Yuval Tassa, and Piotr Trochim.
- My colleagues and collaborators: Boris, Fabio, Hany, João, Joe, Pascal, Samuele, Svenja, and the other members of IAS.
- My master students: Christian, Kai, Kay, Janosch, Johannes, and Daniel.

Contents

Acronyms

cFVI	Continuous Fitted Value Iteration
DeLaN	Deep Lagrangian Networks
DiffNEA	Differentiable Newton-Euler Algorithm
FVI	Fitted Value Iteration
HJB	Hamilton-Jacobi-Bellman
HJI	Hamilton-Jacobi-Isaacs
HNN	Hamiltonian Neural Networks
MLP	Multi-Layer Perceptron
MSE	Mean squared error
nMSE	Normalized mean squared error
ODE	Ordinary differential equation
ProMP	Probabilistic Movement Primitive
rFVI	Robust Fitted Value Iteration
RL	Reinforcement learning
Sim2Real	Simulation to real
VPT	Valid prediction time

Chapter 1
Introduction

A fundamental problem of robotics is how can one program a robot to perform a task with its limited embodiment? Classical robotics solves this problem by carefully engineering separate modules for each part. These modules are connected through an interpretable interface. For example, one module perceives the environment and estimates the engineered system state. The next module uses the state together with a simulator to plan future actions. The subsequent controller uses the manually tuned gains to translate the plan into motor torques. Using this approach, complex systems have been build that enable humanoids to perform backflips [6], rovers driving on Mars, and manipulators assembling cars. The main advantage of classical robotics is that individual modules can be reused for different tasks and embodiments and that the resulting robot behaviors are comprehensible. The main disadvantage of this technique is that these modules must be manually developed, arranged, and tuned for each task. Therefore, engineering these systems is labor-intensive and requires expert knowledge. For more complex tasks, unstructured environments, and unstructured observations, the associated complexity for each module increases and can hardly be developed manually. Therefore, classical robotics becomes financially unfeasible to impossible in unstructured settings. Due to this limitation, robots have not yet left the factory floors and have not entered the everyday life in our households.

Robot learning proposes to trade-off manual engineering for data. Using data should enable robot programs that apply to more complex environments and reduce the required labor for developing the robot program. One approach is to take the data versus engineering trade-off to an extreme. In this brute-force paradigm, one solely relies on data and replaces all modules of classical robotics with interconnected deep networks. The environment and task are assumed to be a black box and unknown. The optimal network parameters are learned end-to-end using reinforcement learning or self-supervised losses that predict the future. The pioneer of reinforcement learning Rich Sutton argues that incorporating prior knowledge is a fallacy as these inductive biases have historically led to poor performance [25]. Such deep learning approaches have achieved astonishing results in simulation using only

M. Lutter, *Inductive Biases in Machine Learning for Robotics and Control*,
Springer Tracts in Advanced Robotics 156,
https://doi.org/10.1007/978-3-031-37832-4_1

pixels [11–13] and researchers have started to transfer these results to physical systems more recently [3, 10].

The advantage of this black box method with end-to-end learning is its generic applicability and simplicity. One does not require the domain knowledge to select, compose and tune the different modules. The disadvantage is that these methods require a lot of data, do not extrapolate, and can overfit to spurious correlations. The amount of required data is prohibitive for robotics as the data must be generated on the physical system for each task and embodiment. Unlike computer vision or natural language processing, the problem is also not static. As the robot operates, the dynamics and optimal policy change. Even if a large dataset could be collected, it would not be sufficient as the dataset would not capture the changes over time. Deep networks perform well for interpolation of high-dimensional spaces but are inherently local. Therefore, these learned approximations are only applicable to the training domain and do not extrapolate beyond. The overparameterization of deep networks lets these approximations easily overfit to spurious correlations observed within the data. This problem is specifically pronounced for reinforcement learning. In this paradigm, overfitting can lead to undesired solutions. The hand designed reward function, the combination of system dynamics and reward function, the used approximations for policy and value function, as well as other optimization biases frequently introduce unintentional local maximums. For example, robots have learned to move using their neck instead of their legs [15] or using somersaults [26].

In this book, we want to show that these two approaches of classical engineering and black-box deep networks are not mutually exclusive. One can transfer insights from classical robotics to the black box deep networks to obtain better learning algorithms for robotics and control. Incorporating the known structure as inductive bias improves the generalization, reduces the sample complexity, and prevents overfitting to spurious correlations as the incorporated structure reduces the model capacity and prevents overfitting. Therefore, the main research question of this book is:

How can one combine existing knowledge and data-driven deep learning methods to learn models and policies applicable to physical robots?

To answer this question, we show that prior knowledge can be beneficial for learning robotic tasks and propose three algorithms that combine domain knowledge and learning. First, we reinterpret the classical system identification approach using the machine learning toolset. Subsequently, we propose two approaches that integrate generic inductive biases into deep networks to learn energy-conserving dynamics models and optimal policies that are robust w.r.t. varying dynamics.

Using prior knowledge for robot learning is not novel. Most of the classical robot learning approaches kept the modular structure and utilized domain knowledge to derive structured representations that are compatible with learning techniques. The proposed representations had very few learnable parameters to learn with only a handful of trials [1, 2, 17–19]. Especially before the deep learning hype, linear representations with engineered features were the pre-dominant representation for robot learning. These features were carefully designed to contain the relevant information about the embodiment and task because the success and failure did depend on the features. For example, researchers rearranged the inverse dynamics equations of open-loop kinematic chains to derive features that simplify the inverse dynamics computation to a linear representation [4, 7, 16, 20]. Others proposed linear motion representations that use time-dependent basis functions as features [14, 21, 22, 24].

In contrast to these existing robot learning techniques, we show that this combination can cover the complete spectrum ranging from the highly constrained classical robotics modules to generic deep networks. Furthermore, the proposed inductive biases are not necessarily specific to the task or embodiment but apply to a large class of systems. Therefore, some of the inductive biases are generic and do not incapacitate the learning algorithms.

1.1 Contribution

To show that one can combine data-driven learning and prior knowledge from physics, control, and robotics, we present three different algorithms. These algorithms can either learn a dynamics model or an optimal policy. The proposed combinations either reinterpret classical methods or are generic inductive biases for black-box deep networks. For example, the differentiable Newton–Euler algorithm extends the classical system identification technique to more dynamical systems with the machine learning toolset. Deep Lagrangian Networks only rely on deep networks and use the incorporated structure to guarantee the conservation of energy.

The presented algorithms are an initial proof of concept of this combination and showcase the potential of data-driven learning with inductive biases. These algorithms do not solve all problems of black-box deep learning based approaches or classical engineering approaches and have their limitations. Future work is needed to scale these combinations to more challenging tasks and observe how far this combination can take us. Nevertheless, these algorithms are an important first step to showcase the potential of inductive biases for black-box approaches. In the following, we briefly introduce the three algorithms and their contributions. The corresponding related work and discussion about the limitations are provided in the respective chapters.

1.1.1 Differentiable Newton–Euler Algorithm

The Differentiable Newton–Euler Algorithm (DiffNEA) reinterprets the classical system identification technique of identifying the dynamics parameters from data. Instead of using linear regression to obtain the parameters [4, 7, 16, 20], this approach leverages automatic differentiation, virtual parameters, and gradient-based optimization. Therefore, DiffNEA guarantees physically consistent parameters whereas the classical methods do not. Furthermore, we show that this approach applies to a wider class of dynamical systems. This algorithm learns the parameters of various friction

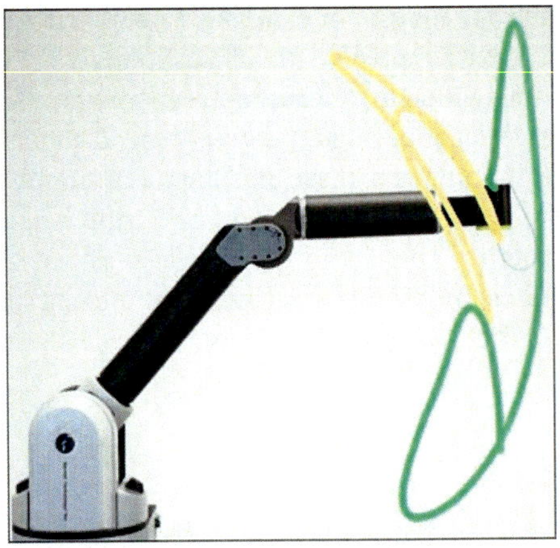

Fig. 1.1 The Ball in Cup task swinging a ball on a string into the cup. The robot movement is optimized using model-based RL. The learned DiffNEA model enables the successful transfer to the physical system

models and systems that contain non-rigid links and non-holonomic constraints. The extensive experimental evaluation shows, that this system identification approach learns accurate dynamics models of real-world systems. Especially in the offline reinforcement learning experiments, the proposed approach excels as the learned dynamics model must extrapolate. For Ball in a Cup (Fig. 1.1), the DiffNEA model solves the task using model-based offline reinforcement learning on the physical system. Therefore, this approach learns accurate parameters of the cup and the string. The black box model learning approaches fail on this task in simulation and on the physical system despite using more data. The black box models can be easily exploited by reinforcement learning (RL) reinforcement learning and learn random movements due to the exploitation.

1.1.2 Deep Lagrangian Networks

Deep Lagrangian Networks (DeLaN) combines deep networks with Lagrangian mechanics to learn dynamics models that conserve energy. The proposed approach uses two deep networks to parameterize the potential and kinetic energy of the system. Combining both networks yields the Lagrangian which can be used to compute the forward and inverse model (Fig. 1.2). The optimal network parameters are obtained by minimizing the squared residual of the Euler–Lagrange differential equation. Therefore, the DeLaN model learns the system energy unsupervised. The resulting DeLaN models retain many advantages of classical system identification techniques but do not require any specific knowledge of the individual system as the classical methods. DeLaN models are interpretable, conserve energy, and can be used as a forward, inverse, and energy model. Since our initial introduction of DeLaN many variants of physics-inspired deep networks have been proposed [5, 8, 9, 23, 27]. The experimental evaluation shows that these learned models can be used for real-time

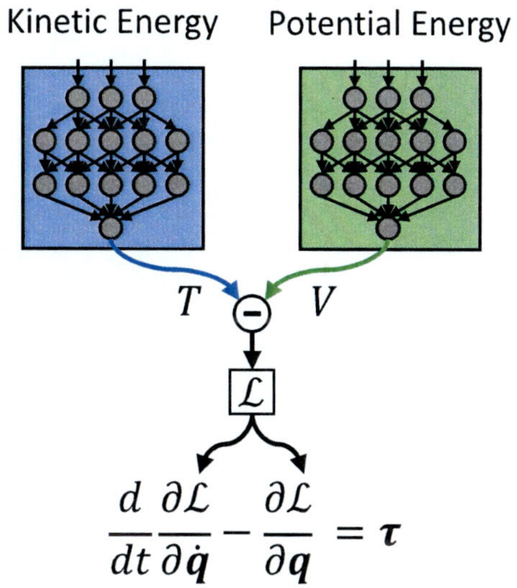

Fig. 1.2 The computational graph of DeLaN. Two networks approximate the Lagrangian $\mathcal{L}$. The Euler–Lagrange equation is used to obtain the dynamics

control of simulated and physical rigid body systems. Compared to standard deep networks, the physics-inspired models learn better models and capture the underlying structure of the dynamics. DeLaN enables energy control that regulates the system energy instead of the joint positions and velocities. Using energy control, DeLaN achieves the swing up of the under-actuated Furuta pendulum and the cartpole. Previously, energy control was not possible using black-box model learning approaches as these approaches cannot learn the system energy.

1.1.3 Robust Fitted Value Iteration

Robust Fitted Value Iteration (rFVI) extends value iteration to continuous actions and the adversarial reinforcement learning problem. We leverage the non-linear control-affine dynamics of many mechanical systems as well as the separable state and action reward of many continuous control problems to derive the optimal policy and optimal adversary in closed form. This analytic expression simplifies the Hamilton–Jacobi–Isaacs differential equation. The resulting equation enables us to solve for the optimal value function using value iteration for continuous actions and states as well as the adversarial case. The control experiments using the Furuta pendulum and cartpole show that rFVI obtains the optimal policy (Fig. 1.3). The robustness sim2real experiments on the physical systems show that the policies successfully achieve the task in the real world. When changing the masses of the pendulum, we observe that rFVI is more robust compared to deep reinforcement learning algorithm and the non-robust version of the algorithm (Fig. 1.2).

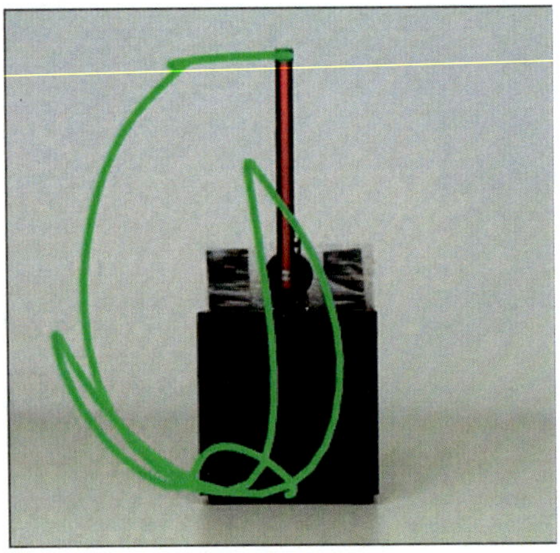

Fig. 1.3 The successful swing-up of the under-actuated Furuta Pendulum using the optimal policy obtained by rFVI

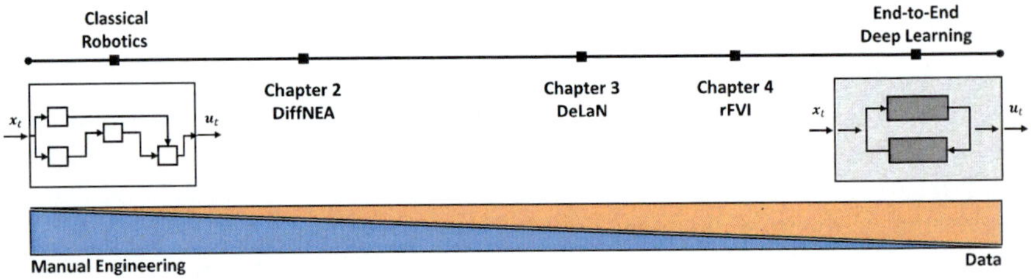

Fig. 1.4 The presented algorithms in Chaps. 2–4 cover the complete spectrum between classical engineering and data driven end-to-end learning with deep networks. Therefore, these algorithm shift the manual engineering versus data trade-off

1.2 Book Outline

The main chapters (Chaps. 2–4) cover the complete spectrum between classical engineering and data driven end-to-end deep learning (Fig. 1.4). Each chapter introduces a specific research question, the proposed solution, and the performed experiments. The concluding Chap. 5 summarizes the book and states the conclusions. The topic of the individual chapters is:

This chapter introduces the idea of combining existing knowledge with data-driven learning. In addition, the differences to existing approaches are highlighted.

Chapter 2 presents the differentiable Newton–Euler algorithm that can identify the system parameters of a mechanical system. The algorithm is evaluated by applying the system identification technique to trajectory prediction and model-based reinforcement learning of physical systems.

Chapter 3 proposes physics-inspired deep networks for learning dynamics models. Specifically, we present Deep Lagrangian Networks (DeLaN) that combine

Lagrangian mechanics with deep learning. The learned dynamics models are evaluated using inverse dynamics control as well as energy control on the physical system.

Chapter 4 extends value iteration to domains with continuous actions and the adversarial reinforcement learning problem. The resulting dynamic programming approach, robust fitted value iteration, is applied to standard control benchmark using simulation to real experiments.

Chapter 5 summarizes the book and presents future work. We discuss the open problems of incorporating inductive biases into machine learning algorithms for robotics and control and propose potential solutions.

References

1. Aboaf, E.W., Atkeson, C.G., Reinkensmeyer, D.J.: Task-level robot learning. In: International Conference on Robotics and Automation (ICRA) (1988)
2. Aboaf, E.W., Drucker, S.M., Atkeson, C.G.: Task-level robot learning: juggling a tennis ball more accurately. In: International Conference on Robotics and Automation (ICRA) (1989)
3. Akkaya, I., Andrychowicz, M., Chociej, M., Litwin, M., McGrew, B., Petron, A., Paino, A., Plappert, M., Powell, G., Ribas, R., et al.: Solving rubik's cube with a robot hand (2019). arXiv:1910.07113
4. Atkeson, C.G., An, C.H., Hollerbach, J.M.: Estimation of inertial parameters of manipulator loads and links. Int. J. Robot. Res. (1986)
5. Cranmer, M., Greydanus, S., Hoyer, S., Battaglia, P., Spergel, D., Ho, S.: Lagrangian neural networks (2020). arXiv:2003.04630
6. Dynamics, B.: What's new, Atlas? Youtube (2017). https://www.youtube.com/watch?v=fRj34o4hN4I
7. Gautier, M.: Identification of robots dynamics. IFAC Proc. Vol. (1986)
8. Greydanus, S., Dzamba, M., Yosinski, J.: Hamiltonian neural networks. In: Advances in Neural Information Processing Systems (NeuRIPS) (2019)
9. Gupta, J.K., Menda, K., Manchester, Z., Kochenderfer, M.J.: A general framework for structured learning of mechanical systems (2019). arXiv:1902.08705
10. Haarnoja, T., Zhou, A., Abbeel, P., Levine, S.: Soft actor-critic: Off-policy maximum entropy deep reinforcement learning with a stochastic actor. In: International Conference on Machine Learning (ICML) (2018)
11. Hafner, D., Lillicrap, T., Ba, J., Norouzi, M.: Dream to control: Learning behaviors by latent imagination (2019). arXiv:1912.01603
12. Hafner, D., Lillicrap, T., Fischer, I., Villegas, R., Ha, D., Lee, H., Davidson, J.: Learning latent dynamics for planning from pixels. In: International Conference on Machine Learning (ICML) (2019)
13. Heess, N., Sriram, S., Lemmon, J., Merel, J., Wayne, G., Tassa, Y., Erez, T., Wang, Z., Eslami, S., Riedmiller, M., et al.: Emergence of locomotion behaviours in rich environments (2017). arXiv:1707.02286
14. Ijspeert, A.J., Nakanishi, J., Hoffmann, H., Pastor, P., Schaal, S.: Dynamical movement primitives: learning attractor models for motor behaviors. Neural Comput. (2013)
15. Irpan, A.: Deep reinforcement learning doesn't work yet (2018). https://www.alexirpan.com/2018/02/14/rl-hard.html
16. Khosla, P.K., Kanade, T.: Parameter identification of robot dynamics. In: Conference on Decision and Control (CDC) (1985)
17. Kober, J., Glisson, M., Mistry, M.: Playing catch and juggling with a humanoid robot. In: International Conference on Humanoid Robots (Humanoids) (2012)

18. Kober, J., Peters, J.R.: Policy search for motor primitives in robotics. In: Advances in Neural Information Processing Systems (2009)
19. Kormushev, P., Calinon, S., Caldwell, D.G.: Robot motor skill coordination with em-based reinforcement learning. In: International Conference on Intelligent Robots and Systems (IROS) (2010)
20. Mukerjee, A., Ballard, D.: Self-calibration in robot manipulators. In: International Conference on Robotics and Automation (ICRA) (1985)
21. Paraschos, A., Daniel, C., Peters, J., Neumann, G.: Using probabilistic movement primitives in robotics. Auton. Robots (2018)
22. Paraschos, A., Daniel, C., Peters, J.R., Neumann, G.: Probabilistic movement primitives. In: Advances in Neural Information Processing Systems (NeurIPS) (2013)
23. Saemundsson, S., Terenin, A., Hofmann, K., Deisenroth, M.: Variational integrator networks for physically structured embeddings. In: International Conference on Artificial Intelligence and Statistics (Aistats) (2020)
24. Schaal, S.: Dynamic movement primitives-a framework for motor control in humans and humanoid robotics. In: Adaptive Motion of Animals and Machines (2006)
25. Sutton, R.: The bitter lesson (2019). http://www.incompleteideas.net/IncIdeas/BitterLesson.html
26. Zhang, B., Rajan, R., Pineda, L., Lambert, N., Biedenkapp, A., Chua, K., Hutter, F., Calandra, R.: On the importance of hyperparameter optimization for model-based reinforcement learning. In: International Conference on Artificial Intelligence and Statistics (Aistats) (2021)
27. Zhong, Y.D., Dey, B., Chakraborty, A.: Symplectic ode-net: learning hamiltonian dynamics with control (2019). arXiv:1909.12077

Chapter 2
A Differentiable Newton–Euler Algorithm for Real-World Robotics

2.1 Introduction

The identification of dynamical systems from data is a powerful tool in robotics [3]. Learned analytic models may be used for control synthesis and can be utilized for gravitational and inertial compensation [75]. When used as simulators, these reduce the sample complexity of data-driven control methods such as RL [11, 15, 28, 46]. For these control applications, where out-of-distribution prediction is typically required, the ability to generalize beyond the acquired data is critical. Any modeling error may be exploited by the data-driven control methods, and such exploitation may result in failure on the physical system. To ensure sufficient out-of-sample generalization, the model's hypothesis space is an important consideration. Ideally, this space should be defined such that only plausible trajectories, that are physically consistent and have bounded energy, are generated.

Standard black-box models such as deep networks or Gaussian processes have a broad hypothesis space and can model arbitrary dynamical systems with high fidelity. Therefore, these black-box models have been frequently used for model learning for control [56–58] as well as model-based reinforcement learning [11, 15, 28, 46]. However, one disadvantage of these black-box models is that the learned approximation is valid locally and not physically consistent. Therefore, these models do not generalize out-of-distribution and can generate trajectories with unbounded energy. Only white-box models [4, 7, 18–22, 24, 36, 40, 42–44, 51, 53, 62, 73, 74, 76, 77, 79, 80, 82, 84], which are derived from first principles and infer the physical parameters of the analytic equations of motion, can guarantee out-of-sample generalization as these models are valid globally. While these combinations of physics with data-driven learning can obtain more robust representations, the usage of white-box models commonly reduces model accuracy compared to black-box methods and has been mainly applied to rigid body systems with holonomic constraints [4, 7, 18–22, 24, 36, 40, 42–44, 51, 53, 62, 73, 74, 76, 77, 79, 80, 82, 84]. The white-box models

M. Lutter, *Inductive Biases in Machine Learning for Robotics and Control*,
Springer Tracts in Advanced Robotics 156,
https://doi.org/10.1007/978-3-031-37832-4_2

obtain a lower accuracy as the Newtonian-, Lagrangian- and Hamiltonian mechanics used to derive the equations of motion typically cannot describe the complex non-linear phenomena of friction with high fidelity. Therefore, these models commonly underfit for physical systems.

In this chapter, we show that learning the system parameters by utilizing differentiable simulators can alleviate many shortcomings of classical white-box models. When the equations of motion are differentiable, we show that one can include complex friction models and apply this approach to systems with non-holonomic constraints. We present a Differentiable Newton–Euler Algorithm (DiffNEA) [49, 50, 76] and combine this approach with various actuator models ranging from white-box friction models to deep network friction models. In the experiments, we apply this differentiable simulator to the identification of multi-body robot dynamics. We benchmark these models on the simulated and physical Furuta Pendulum and Cartpole. The experiments show that DiffNEA models with energy-bounded black-box actuator models yield non-divergent trajectories and improve the predicted rollouts. Other black-box models that do not guarantee passive friction models are susceptible to learn dynamics that generate energy and lead to divergence.

For systems with non-holonomic constraints, we show that these constraints can be added to the optimization as additional penalties. Solving the resulting optimization problem with gradient-based optimization enables the identification of the physical parameters of the non-holonomic constraints. In the experiments, we apply this technique to solve ball in a cup with offline model-based reinforcement learning on the physical system. This task is especially challenging as one needs to learn the string dynamics and obtain a model that cannot be exploited by the reinforcement learning agent. The DiffNEA model can learn an accurate and robust model that cannot be exploited. The optimal solution obtained using the DiffNEA model can be transferred to the physical system and successfully achieves the task for multiple string lengths. In contrast, black-box models can be easily exploited by the RL agent and fail when transferred to the physical system.

The main objective of this chapter is to address two fundamental research questions: (i) Can accurate dynamics models be obtained on the physical system even when the systems include stiction or non-holonomic constraints? (ii) When do differentiable white-box models significantly outperform black-box models on the physical system? To answer these research questions, we utilize the DiffNEA approach with differentiable simulators, first-order optimization and nonlinear friction models. For the first time, we show that this approach can identify dynamics models of dynamical systems that include non-holonomic constraints and nonlinear friction. In addition, we empirically show that such white-box models excel compared to black-box models when used for model-based reinforcement learning. Due to the inherent generalization of white-box models beyond the training domain, this model category minimizes model exploitation of the policy optimization. In the ball in cup experiments, the white-box model enables solving the tasks while the black-box model is easily exploited and converges to random movements. Therefore, white-box models are especially beneficial when generalization is essential.

2.1.1 Naming Convention

In this chapter, we use the term Differentiable Newton Euler Algorithm (DiffNEA) to indicate a differentiable simulator based on the equations of motions derived using the Newton Euler equation. We specifically highlight the differentiability as this characteristic enables the identification of the physical parameters using gradient-based optimization. The term was originally coined by [76]. The equations of motions for the considered systems are differentiable by default. Therefore, the differentiable in DiffNEA does not imply that the equations of motions are transformed to make them differentiable. Moreover, we generally refer to obtaining the parameters of white-box models as 'identifying' the model, while the parameters of black-box models are 'learned'. Depending on the research community both terms are used interchangeably. If we refer to a generic dynamics model we use these words interchangeably.

2.1.2 Outline

The chapter is structured as follows. First, we summarize the different approaches to learning dynamics models for robotics (Sect. 2.2). Section 2.3 describes DiffNEA. The subsequent experimental section (Sect. 2.4) summarizes the experimental setup and presents the results. Finally, Sect. 2.5 discusses the obtained results and summarizes the results of the chapter.

2.2 Dynamics Model Representations

Model learning, or system identification [3], aims to infer the parameters $\boldsymbol{\theta}$ of the system dynamics from data containing the system state $\boldsymbol{x}$ and the control signal $\boldsymbol{\tau}$. In the continuous time case the dynamics are described by

$$\ddot{\boldsymbol{x}} = \boldsymbol{f}(\boldsymbol{x}, \dot{\boldsymbol{x}}, \boldsymbol{\tau}; \boldsymbol{\theta}). \tag{2.1}$$

The optimal parameters $\boldsymbol{\theta}^*$ are commonly obtained by minimizing the error of the forward or inverse dynamics model,

$$\boldsymbol{\theta}^*_{\text{for}} = \arg\min_{\boldsymbol{\theta}} \sum_{i=0}^{N} \|\ddot{\boldsymbol{x}}_i - \hat{\boldsymbol{f}}\,(\boldsymbol{x}_i, \dot{\boldsymbol{x}}_i, \boldsymbol{\tau}_i; \boldsymbol{\theta})\,\|^2, \tag{2.2}$$

$$\boldsymbol{\theta}^*_{\text{inv}} = \arg\min_{\boldsymbol{\theta}} \sum_{i=0}^{N} \|\boldsymbol{\tau}_i - \hat{\boldsymbol{f}}^{-1}\,(\boldsymbol{x}_i, \dot{\boldsymbol{x}}_i, \ddot{\boldsymbol{x}}_i; \boldsymbol{\theta})\,\|^2. \tag{2.3}$$

Depending on the chosen representation for f, the model hypotheses spaces and the optimization method changes. Generally one can differentiate between black-box models and white-box models. Recently there are also have been various gray-box models [2, 34, 49, 55] to bridge the gap and combine parts of both categories.

2.2.1 Black-Box Models

These models use generic function approximators f and the corresponding abstract parameters $\boldsymbol{\theta}$ to represent the dynamics. Within the model learning literature many function approximation technique has been applied, including locally linear models [5, 72], Gaussian processes [39, 55, 58], deep- [11, 28, 46] and graph networks [70, 71]. These approximators can fit arbitrary and complex dynamics with high fidelity but have an undefined behavior outside the training distribution and might be physically unplausible even in the training domain. Due to the local nature of the representation, the behavior is only well defined on the training domain and hence the learned models do not extrapolate well. Furthermore, these models can learn implausible systems violating fundamental physics laws such as energy conservation. Furthermore, these models can only learn either the forward or inverse models and are not interpretable. Therefore, only the input-output relation can be computed but no additional information such as the system energies or momentum. Only recently deep networks have been augmented with knowledge from physics to constrain network representations to be physically plausible on the training domain and interpretable [13, 26, 27, 47, 48, 69, 85]. However, the behavior outside the training domains remains unknown.

2.2.2 White-Box Models

These models use the analytical equations of motions to formalize the hypotheses space of f and the interpretable physical parameters such as mass, inertia, or length as parameters $\boldsymbol{\theta}$. Commonly the equations of motion are derived using either Newtonian, Lagrangian, or Hamiltonian mechanics. Therefore, white-box models are limited to describe the phenomena incorporated within the equations of motions but generalize to unseen state regions as the parameters are global. The classical approach to obtain the system parameters is by measuring these properties of the disassembled system or estimating them using the CAD software [1]. Instead of measuring the parameters, four different groups concurrently proposed to estimate these parameters from data [4, 18, 36, 53]. These papers showed that the recursive Newton–Euler algorithm (RNEA) [17] for rigid-body chain manipulators simplifies to a linear model. Therefore, the parameters can be inferred using linear least squares. The main disadvantage of this approach is that these parameters have additional constraints. For example, the inertial parameters must yield a positive definite inertia matrix and fulfill the triangle inequality [79], the link mass must be positive. However, the identification of the parameters using linear least squares ignores these constraints. Therefore, the obtained dynamics parameters are not necessarily physically consistent.

To obtain physically consistent parameters, many authors have proposed system identification techniques that guarantee consistent parameters [7, 19–22, 24, 40, 42–44, 51, 73, 74, 76, 77, 79, 80, 82, 84]. For example, [77] projected the learned parameters to the set of feasible parameters. References [19, 20] added regularization that ensures that the parameters are close to the CAD values. References [43, 73, 74, 82] derive linear-matrix inequalities to ensure physical consistency. The resulting constrained optimization problem is convex and the global optimal solution can be efficiently obtained via semidefinite programming. References [21, 22] used optimization penalties and global optimization to solve the constrained nonlinear optimization. References [76, 79] used virtual parameters that reparametrize the physical parameters to ensure physical plausibility and solve the resulting optimization using gradient based optimization.

The previous approaches assumed that the equations of motion are analytically known. To avoid this assumption, multiple authors [8, 54, 68] have proposed likelihood-free inference methods to identify the posterior distribution over the parameters. In this case, the equations of motions are abstracted as a black box that only needs to be evaluated. For the standard physics engines, e.g., PyBullet [12] and MuJoCo [78], this assumption is favorable as these simulators only provide an interface to compute the next state and do not provide the derivatives with respect to the current state or the model parameters. Similarly to likelihood-free inference parameter identification for black-box simulators, Jiang et al. [35] proposed to obtain the distribution of the physical parameters using reinforcement learning. These approaches are commonly applied to domain randomization as these techniques automatically tune the randomization for the specific physical system. Therefore, the resulting policies are not too conservative but achieve the simulation to reality transfer.

In this chapter, we assume that the equations of motion are known and differentiable. Therefore, we use a gradient-based approach to infer the optimal parameters. We utilize the recent availability of automatic differentiation (AD) [67] frameworks to compute the gradient w.r.t. to the parameters analytically using backpropagation. Furthermore, we use the virtual parametrization proposed by [76, 79] to guarantee that the physical parameters are plausible without enforcing additional constraints within the optimization.

Algorithm 2.1 The Articulated Body Algorithm computing the joint accelerations for a kinematic tree using Lie algebra [37]

Input: Joint Position $\boldsymbol{q}$, Joint Velocity $\dot{\boldsymbol{q}}$, Torque $\boldsymbol{\tau}$

Output: Joint Acceleration $\ddot{\boldsymbol{q}}_i$

for $i = 1$ to n **do**

Compute forward kinematics along the chain

$\bar{\boldsymbol{v}}_i = Ad_{T_{\lambda,i}^{-1}} \bar{\boldsymbol{v}}_\lambda + \boldsymbol{s}_i \dot{\boldsymbol{q}}_i$

$\bar{\boldsymbol{\eta}}_i = ad_{\bar{\boldsymbol{v}}_i} \boldsymbol{s}_i \, \dot{\boldsymbol{q}}_i$

end for

for $i = n$ to 1 **do**

Compute lumped inertia for each body

$\hat{\bar{\boldsymbol{M}}}_{i:n} = \bar{\boldsymbol{M}}_i + \sum_{k \in \mu} Ad^*_{T_{i,k}^{-1}} \bar{\boldsymbol{\Pi}}_k Ad_{T_{i,k}^{-1}}$

Compute bias forces for each link

$\bar{\boldsymbol{f}}_{b,i} = -ad^*_{\bar{\boldsymbol{v}}_i} \bar{\boldsymbol{M}}_{i:n} \bar{\boldsymbol{v}}_i + \sum_{k \in \mu} Ad^*_{T_{i,k}^{-1}} \left(\bar{\boldsymbol{f}}_{b,k} + \bar{\boldsymbol{\beta}}_k \right)$

$\bar{\Psi}_i = \left(\boldsymbol{s}_i^\top \hat{\bar{\boldsymbol{M}}}_{i:n} \, \boldsymbol{s}_i \right)^{-1}$

$\bar{\boldsymbol{\Pi}}_i = \hat{\bar{\boldsymbol{M}}}_{i:n} - \hat{\bar{\boldsymbol{M}}}_{i:n} \boldsymbol{s}_i \bar{\Psi}_i \boldsymbol{s}_i^\top \hat{\bar{\boldsymbol{M}}}_{i:n}$

$\bar{\boldsymbol{\beta}}_i = \hat{\bar{\boldsymbol{M}}}_{i:n} \left(\bar{\boldsymbol{\eta}}_i + \boldsymbol{s}_i \bar{\Psi}_i \left(u_i - \boldsymbol{s}_i^\top \left(\hat{\bar{\boldsymbol{M}}}_{i:n} \bar{\boldsymbol{\eta}}_i + \bar{\boldsymbol{f}}_{b,i} \right) \right) \right)$

end for

for $i = 1$ to n **do**

Compute acceleration using Newton–Euler

$\ddot{\boldsymbol{q}}_i = \bar{\Psi}_i \left(\boldsymbol{\tau}_i - \boldsymbol{s}_i^\top \left(\hat{\bar{\boldsymbol{M}}}_{i:n} \left(Ad_{T_{\lambda,i}^{-1}} \bar{\boldsymbol{a}}_\lambda + \bar{\eta}_i \right) - \bar{\boldsymbol{f}}_{b,i} \right) \right)$

$\bar{\boldsymbol{a}}_i = Ad_{T_{\lambda,i}^{-1}} \bar{\boldsymbol{a}}_\lambda + \boldsymbol{s}_i \ddot{\boldsymbol{q}}_i + \bar{\boldsymbol{\eta}}_i$

$\bar{\boldsymbol{f}}_i = \hat{\bar{\boldsymbol{M}}}_{i:n} \bar{\boldsymbol{a}}_i + \bar{\boldsymbol{f}}_{b,i}$

end for

2.2.3 Differentiable Simulators

As we assume that the equations of motion are differentiable, the proposed system identification approach using gradient-based optimization is closely related to differentiable simulators. Recently various approaches to differentiable simulation [6, 14, 23, 29–31, 33, 83] has been proposed to enable system identification via gradients and policy optimization using backprop through time [52]. The main problem of differentiable simulators is differentiating through the contact and the constraint force solver computing $\boldsymbol{f}_c$. The equations of motion of articulated rigid bodies without contacts only require linear algebra operations and hence, are differentiable by default. To differentiate through the contacts, various approaches have been proposed. For example, Belbute-Peres et al. [6] and Heiden et al. [30] describe a method to differentiate through the common LCP solver of simulators, Degrave et al. [14] utilize impulse-based contacts based on constraint violation to enable gradient computation using automatic differentiation and avoid solving the LCP, Geilinger et al. [23] describe a smoothed contact model to enable the differentiation and Hu et al. [33] describe a continuous collision resolution approach to improve the gradient computation. A great in-depth description of the different approaches to differentiable simulation and their advantages is provided in [83].

In this chapter, we mainly only simulate articulated rigid bodies that are differentiable by default. For ball in a cup, the constraint forces originating from the non-holonomic constraints can be computed in closed form. Therefore, we can easily differentiate through these analytic expressions of the constraint forces and do not need to rely on more complex approaches presented in the literature.

2.3 Differentiable Newton–Euler Algorithm

In this section, we describe the differentiable system identification technique that is based on the Newton–Euler algorithm in terms of the elegant Lie algebra formulation [10, 62] in addition to the technical report [37]. First, we describe the identification approach for systems with holonomic constraints, i.e., kinematic trees (Sect. 2.3.1). Afterwards, we describe the virtual parameters that enable physically plausible system parameters (Sect. 2.3.2) and extend them to systems with non-holonomic constraints (Sect. 2.3.3). Finally, the different actuators are introduced in Sect. 2.3.4.

2.3.1 Rigid-Body Physics and Holonomic Constraints

For rigid-body systems with holonomic constraints, the system dynamics can be expressed analytically in maximal coordinates $\boldsymbol{x}$, i.e., task space, and reduced coordinates $\boldsymbol{q}$, i.e., joint space. If expressed using maximal coordinates, the dynamics is a constrained problem with the holonomic constraints $g(\cdot)$. For the reduced coordinates, the dynamics are reparametrized such that the constraints are always fulfilled and the dynamics are unconstrained. Mathematically these dynamics are described by

$$\ddot{\boldsymbol{x}} = f(\boldsymbol{x}, \dot{\boldsymbol{x}}, \boldsymbol{\tau}; \boldsymbol{\theta}) \quad \text{s.t.} \quad g(\boldsymbol{x}; \boldsymbol{\theta}) = 0, \tag{2.4}$$

$$\Rightarrow \ddot{\boldsymbol{q}} = f(\boldsymbol{q}, \dot{\boldsymbol{q}}, \boldsymbol{\tau}; \boldsymbol{\theta}). \tag{2.5}$$

For model learning of such systems one commonly exploits the reduced coordinate formulation and minimizes the squared loss of the forward or inverse model. For kinematic trees the forward dynamics $f(\cdot)$ can be easily computed using the articulated body algorithm (ABA, Algorithm 2.1) and the inverse dynamics $f^{-1}(\cdot)$ via the recursive Newton–Euler algorithm (RNEA, Algorithm 2.2) [17]. Both algorithms are inherently differentiable and one can solve the optimization problem of Eq. 2.2 using backpropagation.

Algorithm 2.2 The Recursive Newton–Euler Algorithm computing the input torque for a kinematic tree in terms of Lie algebra [37]

Input: Position $\boldsymbol{q}$, Velocity $\dot{\boldsymbol{q}}$, Acceleration $\ddot{\boldsymbol{q}}$

Output: Torque $\boldsymbol{\tau}$

for $i = 1$ to n **do**

// Forward Kinematics

$\bar{\boldsymbol{v}}_i = Ad_{T^{-1}_{\lambda,i}} \bar{\boldsymbol{v}}_\lambda + \boldsymbol{s}_i \dot{\boldsymbol{q}}_i$

$\bar{\boldsymbol{a}}_i = Ad_{T^{-1}_{\lambda,i}} \bar{\boldsymbol{a}}_\lambda + ad_{\bar{\boldsymbol{v}}_i} \boldsymbol{s}_i \, \dot{\boldsymbol{q}}_i + \boldsymbol{s}_i \, \ddot{\boldsymbol{q}}_i$

end for

for $i = n$ to 1 **do**

// Newton–Euler Equation

$\bar{\boldsymbol{f}}_i = \bar{\boldsymbol{M}}_i \bar{\boldsymbol{a}}_i - ad^*_{\bar{\boldsymbol{v}}_i} \bar{\boldsymbol{M}}_i \bar{\boldsymbol{v}}_i + \sum_{k \in \mu} Ad^*_{T^{-1}_{i,k}} \bar{\boldsymbol{f}}_k$

$\boldsymbol{\tau}_i = \boldsymbol{s}_i^\top \bar{\boldsymbol{f}}_i$

end for

In this implementation, we use the Lie formulations of ABA and RNEA [37] for compact and intuitive compared to the initial derivations by [4, 17]. ABA and RNEA propagate velocities and accelerations from the kinematic root to the leaves and the forces and impulses from the leaves to the root. This propagation along the chain can be easily expressed in Lie algebra by

$$\bar{\boldsymbol{v}}_j = \mathrm{Ad}_{\boldsymbol{T}_{j,i}} \bar{\boldsymbol{v}}_i, \qquad \bar{\boldsymbol{a}}_j = \mathrm{Ad}_{\boldsymbol{T}_{j,i}} \bar{\boldsymbol{a}}_i, \tag{2.6}$$

$$\bar{\boldsymbol{l}}_j = \mathrm{Ad}^\top_{\boldsymbol{T}_{j,i}} \bar{\boldsymbol{l}}_i, \qquad \bar{\boldsymbol{f}}_j = \mathrm{Ad}^\top_{\boldsymbol{T}_{j,i}} \bar{\boldsymbol{f}}_i. \tag{2.7}$$

with the generalized velocities $\bar{\boldsymbol{v}}$, accelerations $\bar{\boldsymbol{a}}$, forces $\bar{\boldsymbol{f}}$, momentum $\bar{\boldsymbol{l}}$ and the adjoint transform $\mathrm{Ad}_{\boldsymbol{T}_{j,i}}$ from the ith to the jth link. The generalized entities noted by $\bar{\cdot}$ combine the linear and rotational components, e.g., $\bar{\boldsymbol{v}} = [\boldsymbol{v}, \boldsymbol{\omega}]$ with the linear velocity $\boldsymbol{v}$ and the rotational velocity $\boldsymbol{\omega}$. The Newton–Euler equation is described by

$$\bar{f}_{\text{net}} = \bar{M}\bar{a} - \text{ad}^*_{\bar{v}}\bar{M}\bar{v},$$
$$\text{ad}^*_{\bar{v}} = \begin{bmatrix} [\omega] & \mathbf{0} \\ [v] & [\omega] \end{bmatrix}, \quad \bar{M} = \begin{bmatrix} J & m[p_m] \\ m[p_m]^\top & mI \end{bmatrix},$$

with the inertia matrix J, the link mass m, the center of mass offset p_m. Combining this message passing with the Newton–Euler equation enables a compact formulation of RNEA and ABA.

2.3.2 *Virtual Physical Parameters*

To obtain the optimal parameters from data, one cannot simply minimize the mean squared error of the forward or inverse model as these parameters have additional constraints. For example, the transformation matrix T must be a homogeneous transformation (i.e., $T \in \text{SE}(3)$), the inertias J must comply with the parallel axis theorem, and the masses m must be positive. To obtain physically plausible parameters with gradient-based optimization without additional constraints, we reparametrize these physical parameters with *virtual* parameters. These virtual parameters are *unrestricted* and yield a physically plausible for all values [76, 79, 82]. For example, we optimize the square root of the mass to always obtain a positive mass.

2.3.2.1 Kinematics

The transformation $T(q)$ between two links depends on the link length, the joint position and the joint constraint connecting the two links. We decompose this transformation $T(q) = T_O T_q(q)$ into a fixed transform T_O and variable transform $T_q(q_i)$. The fixed transform describes the the distance and rotation between two joints and is parametrized by the translation vector p_k and the RPY Euler angles $\theta_R = [\phi_x, \phi_y, \phi_z]^\top$. The transformation is then described by

$$T_O = \begin{bmatrix} R_z(\phi_z)R_y(\phi_y)R_x(\phi_x) & p_k \\ 0 & 1 \end{bmatrix}, \tag{2.8}$$

where $R_a(\phi)$ denotes the rotation matrix corresponding to the rotation by ϕ about axis a using the right-hand rule. Note that the rotation matrices about the elementary axis only depend on θ_R through arguments to trigonometric functions. Due to the periodic nature of those functions we obtain a desired unrestricted parameterization. One disadvantage of this parametrization is that the Euler angles can loose a degree of freedom during optimization limiting the representational power (i.e., Gimbal Lock) during gradient-based optimization and cause numerical instabilities. In our experiments we did not observe this problem. Other representations exist that do not exhibit this problem. For example, one could parameterize the rotation matrix using the SO(3) Lie algebra and the corresponding exponential map [76] or a constrained representations such as the rotation matrix or quaternion [79]. In the latter case, case one would need to incorporate these constraints into the optimization.

The variable transform $\boldsymbol{T}_q(q_i)$ describes the joint constraint and joint configuration. For all joints we assume that the variable link axis is aligned with the z-axis. Hence, the transformation matrix and joint motion vector for revolute joints ($\boldsymbol{T}_{q_r}, \boldsymbol{s}_r$) and prismatic joints ($\boldsymbol{T}_{q_p}, \boldsymbol{s}_p$) is described by

$$\boldsymbol{T}_{q_r} = \begin{bmatrix} \boldsymbol{R}_z(q) & \boldsymbol{0} \\ \boldsymbol{0} & 1 \end{bmatrix}, \boldsymbol{T}_{q_p} = \begin{bmatrix} \boldsymbol{0} & q\,\boldsymbol{e}_z \\ \boldsymbol{0} & 1 \end{bmatrix},$$
$$\boldsymbol{s}_r = \begin{bmatrix} \boldsymbol{0} & \boldsymbol{e}_z \end{bmatrix}, \boldsymbol{s}_p = \begin{bmatrix} \boldsymbol{e}_z & \boldsymbol{0} \end{bmatrix},$$

with the z axis unit vector $\boldsymbol{e}_z$. Technically, one can also use fixed joints with $\boldsymbol{T}_q = \mathbf{I}$ but this simply yields an overparameterized model and is not necessary for describing the system dynamics. The complete kinematics parameters of a link are summarized as $\boldsymbol{\theta}_K = \{\boldsymbol{\theta}_R, \boldsymbol{p}_k\}$.

2.3.2.2 Inertias

For physical correctness, the diagonal rotational inertia $\boldsymbol{J}_p = \text{diag}([J_x, J_y, J_z])$ at the body's CoM and around principal axes must be positive definite and need to conform with the triangle inequalities [79], i.e.,

$$J_x \leq J_y + J_z, \quad J_y \leq J_x + J_z, \quad J_z \leq J_x + J_y \,.$$

To allow an unbounded parameterization of the inertia matrix, we introduce the parameter vector $\boldsymbol{\theta}_L = [\theta_{\sqrt{L_1}}, \theta_{\sqrt{L_2}}, \theta_{\sqrt{L_3}}]^\intercal$, where L_i represents the central second moments of mass of the density describing the mass distribution of the rigid body with respect to a principal axis frame. Then rotational inertia is described by

$$\boldsymbol{J}_p = \text{diag}\left(\theta^2_{\sqrt{L_2}} + \theta^2_{\sqrt{L_3}},\ \theta^2_{\sqrt{L_1}} + \theta^2_{\sqrt{L_3}},\ \theta^2_{\sqrt{L_1}} + \theta^2_{\sqrt{L_2}}\right).$$

The rotational inertia is then mapped to the link coordinate frame using the parallel axis theorem described by

$$\boldsymbol{J} = \boldsymbol{R}_J \boldsymbol{J}_p \boldsymbol{R}_J^\intercal + m \left[\boldsymbol{p}_m\right]^\intercal \left[\boldsymbol{p}_m\right], \tag{2.9}$$

with the link mass m and the translation $\boldsymbol{p}_m$ from the coordinate from to the CoM. The fixed affine transformation uses the same parameterization as described in Sect. 2.3.2.1. The mass of the rigid body is parameterized by $\theta_{\sqrt{m}}$ where $m = \theta^2_{\sqrt{m}}$ [76, 77]. Therefore, the inertial parameter required to compute the generalized inertia $\bar{\boldsymbol{M}}$ for each link are described by $\boldsymbol{\theta}_J = \{\boldsymbol{\theta}_L, \boldsymbol{\theta}_R, \theta_{\sqrt{m}}, \boldsymbol{p}_m\}$. Alternatively, a physically plausible inertia matrix $\boldsymbol{J}$ could also represented as a symmetric and positive definite 4×4 matrix. Instead of learning the separate quantities for mass, inertia and center of mass, these quantities can be computed from this 4×4 positive definite matrix [43, 44]. An extensive empirical comparison of the different inertia reparametrizations can be found in [76].

2.3.3 Rigid-Body Physics and Non-holonomic Constraints

For a mechanical system with non-holonomic constraints, the system dynamics cannot be expressed in terms of unconstrained equations with reduced coordinates. For the system

$$\ddot{\boldsymbol{x}} = f(\boldsymbol{x}, \dot{\boldsymbol{x}}, \boldsymbol{u}; \boldsymbol{\theta}) \quad \text{s.t.} \quad h(\boldsymbol{x}; \boldsymbol{\theta}) \leq 0, \quad g(\boldsymbol{x}, \dot{\boldsymbol{x}}; \boldsymbol{\theta}) = 0,$$

the constraints are non-holonomic as $h(\cdot)$ is an inequality constraint and $g(\cdot)$ depends on the velocity. Inextensible strings are an example for systems with inequality constraint, while the bicycle is a system with velocity dependent constraints. For such systems, one cannot optimize the unconstrained problem directly but must identify parameters that explain the data and adhere to the constraints.

The dynamics of the constrained rigid body system can be described by the Newton–Euler equation,

$$\bar{\boldsymbol{f}}_{\text{net}} = \bar{\boldsymbol{f}}_g + \bar{\boldsymbol{f}}_c + \bar{\boldsymbol{f}}_{\boldsymbol{u}} = \bar{\boldsymbol{M}}\bar{\boldsymbol{a}} - \text{ad}^*_{\bar{\boldsymbol{v}}}\bar{\boldsymbol{M}}\bar{\boldsymbol{v}}, \tag{2.10}$$

$$\Rightarrow \bar{\boldsymbol{a}} = \bar{\boldsymbol{M}}^{-1}\left(\bar{\boldsymbol{f}}_g + \bar{\boldsymbol{f}}_c + \bar{\boldsymbol{f}}_{\boldsymbol{u}} + \text{ad}^*_{\bar{\boldsymbol{v}}}\bar{\boldsymbol{M}}\bar{\boldsymbol{v}}\right), \tag{2.11}$$

where the net force $\bar{\boldsymbol{f}}_{\text{net}}$ contains the gravitational force $\bar{\boldsymbol{f}}_g$, the constraint force $\bar{\boldsymbol{f}}_c$ and the control force $\bar{\boldsymbol{f}}_{\boldsymbol{u}}$. If one can differentiate the constraint solver computing the constraint force w.r.t. to the parameters, one can identify the parameters $\boldsymbol{\theta}$ via gradient descent. This optimization problem can be described by

$$\min_{\boldsymbol{\theta}} \sum_{i=0}^{N} \left| \bar{\boldsymbol{a}}_i - \bar{\boldsymbol{M}}_{\boldsymbol{\theta}}^{-1}\left(\bar{\boldsymbol{f}}_g(\boldsymbol{\theta}) + \bar{\boldsymbol{f}}_c(\bar{\boldsymbol{x}}_i, \bar{\boldsymbol{v}}_i; \boldsymbol{\theta}) + \bar{\boldsymbol{f}}_{\boldsymbol{u}} + \text{ad}^*_{\bar{\boldsymbol{v}}_i}\bar{\boldsymbol{M}}(\boldsymbol{\theta})\bar{\boldsymbol{v}}_i\right) \right|^2.$$

For the inequality constraint, one can reframe it as an easier equality constraint, by passing the function through a ReLU nonlinearity σ, so $g(\boldsymbol{x}; \boldsymbol{\theta}) = \sigma(h(\boldsymbol{x}; \boldsymbol{\theta})) = 0$. From a practical perspective, the softplus nonlinearity provides a soft relaxation of the nonlinearity for smoother optimization. Since this equality constraint should always be enforced, we can utilize our dynamics to ensure this on the derivative level, so $g(\cdot) = \dot{g}(\cdot) = \ddot{g}(\cdot) = 0$ for the whole trajectory. With this augmentation, the constraint may now be expressed as $\boldsymbol{g}(\boldsymbol{x}, \dot{\boldsymbol{x}}; \boldsymbol{\theta}) = \boldsymbol{0}$. The complete loss is described by

$$\min_{\boldsymbol{\theta}} \sum_{i=0}^{N} \|\bar{\boldsymbol{a}}_i - \boldsymbol{f}(\bar{\boldsymbol{x}}_i, \bar{\boldsymbol{v}}_i, \bar{\boldsymbol{u}}_i; \boldsymbol{\theta})\|^2 + \boldsymbol{g}(\boldsymbol{x}_i, \dot{\boldsymbol{x}}_i; \boldsymbol{\theta})^\top \boldsymbol{\Lambda}\, \boldsymbol{g}(\boldsymbol{x}_i, \dot{\boldsymbol{x}}_i; \boldsymbol{\theta}),$$

with the penalty weighting $\boldsymbol{\Lambda} = \text{diag}\left(\lambda_g, \lambda_{\dot{g}}, \lambda_{\ddot{g}}\right)$.

2.3.4 Actuator Models

All described models simulate rigid body systems that conserve the energy and assume that the actuators apply the desired torque $\boldsymbol{\tau}_d$, i.e., $\boldsymbol{\tau} = \boldsymbol{\tau}_d$. For physical systems, this representation does not capture reality with sufficient fidelity as these systems dissipate energy via friction and the actuators are not ideal, i.e., $\boldsymbol{\tau} \approx \boldsymbol{\tau}_d$. For most articulated kinematic trees, the friction of the joints and actuators dominates compared to the air resistance. To incorporate these phenomena into white-box models, we augment the rigid-body body simulator with an actuator model that incorporates joint friction and transforms the desired torque into the applied torque, i.e., $\boldsymbol{\tau} = f(\boldsymbol{\tau}_d, \boldsymbol{q}, \dot{\boldsymbol{q}})$.

This actuation model can either be white-box model relying on existing friction models or black-box models. We define five different joint independent actuator models that cover the complete spectrum from white-box to black-box model,

$$\begin{aligned}
&\text{Viscous:} && \boldsymbol{\tau} = \boldsymbol{\tau}_d - \boldsymbol{\mu}_v \odot \dot{\boldsymbol{q}},\\
&\text{Stribeck:} && \boldsymbol{\tau} = \boldsymbol{\tau}_d - \sigma(\dot{\boldsymbol{q}}) \odot \left(\boldsymbol{f}_s + \boldsymbol{f}_d \odot \exp\left(-\nu_s \dot{\boldsymbol{q}}^2\right)\right) - \boldsymbol{\mu}_v \odot \dot{\boldsymbol{q}},\\
&\text{NN-Friction:} && \boldsymbol{\tau} = \boldsymbol{\tau}_d - \sigma(\dot{\boldsymbol{q}}) \odot \| f_{\text{NN}}(\boldsymbol{q}, \dot{\boldsymbol{q}}; \psi) \|_1,\\
&\text{NN-Residual:} && \boldsymbol{\tau} = \boldsymbol{\tau}_d + f_{\text{NN}}(\boldsymbol{q}, \dot{\boldsymbol{q}}; \psi),\\
&\text{FF-NN:} && \boldsymbol{\tau} = f_{\text{NN}}(\boldsymbol{\tau}_d, \boldsymbol{q}, \dot{\boldsymbol{q}}; \psi),
\end{aligned}$$

with the deep networks weights ψ and the elementwise multiplication $\odot$. The Viscous and Stribeck actuator models are white-box models that have been proposed within the literature [1, 9, 59, 81]. These models assume that the the motor is ideal as well as that the friction is additive, independent for each joint and does not depend on the robot state. The resulting system dynamics is guaranteed to yield a stable uncontrolled system as these models dissipate energy, i.e., $\dot{E} = \boldsymbol{\tau}^\top \dot{\boldsymbol{q}} \leq 0$. The black-box alternative to these friction model is the NN friction model. This friction model also assumes that applied torque is ideal and guarantees that the system is passive. However, in this case the frictional torque depends on the robot configuration and can represent more complex shapes. To not only model the actuator friction that dissipates energy, the NN residual and FF-NN actuator model can model arbitrary actuator dynamics.

To learn the parameters of the actuator models, we add the parameters of the models to the gradient-based optimization. We regularize the training of the actuator model by adding penalties that prevent the actuator model dynamics to dominate the system dynamics. Similar optimization procedures have been used in [2, 34, 49] to train existing grey-box models that combine deep networks and analytic models.

2.4 Experiments

In the experiments, we apply DiffNEA to three physical systems. We evaluate the long-term predictions of the learned forward models and test whether the models can be used for reinforcement learning. Using this evaluation, we want to answer the following questions:

Q1: Can DiffNEA models learn accurate dynamics models on the physical system even when the systems include stiction or non-holonomic constraints?

Q2: When do DiffNEA models significantly outperform black-box models on the physical system?

To answer these questions we perform two experiments, a trajectory prediction experiment (Sect. 2.4.2) and an offline reinforcement learning experiment (Sect. 2.4.3). In the first experiment, we identify the parameters of two under-actuated systems on different datasets and compare their simulated trajectories. The main objective of this experiment is to compare a large number of different system identification approaches and characterize their similarities and differences. In the second experiment, we learn the ball in a cup dynamics and use the model for offline reinforcement learning. In this experiment, we want to test whether the learned models can be exploited by the reinforcement learning agent. Prior to the experiments, we describe the overarching experimental setup of the different used physical systems, implementations and model variation in Sect. 2.4.1.

2.4.1 Experimental Setup

In this section we describe the used systems, model variations and implementations used in both experiments. The details for the individual experiments are described in the respective sections.

2.4.1.1 Physical Dynamical Systems

We apply the DiffNEA to identify the parameters of the cartpole, the Furuta pendulum, and the Barrett WAM. For all physical systems, only the joint position is directly observed. The velocities and accelerations are computed using finite differences and low-pass filters. These filters are applied offline to use zero-phase shift filters that do not cause a phase shift within the observations.

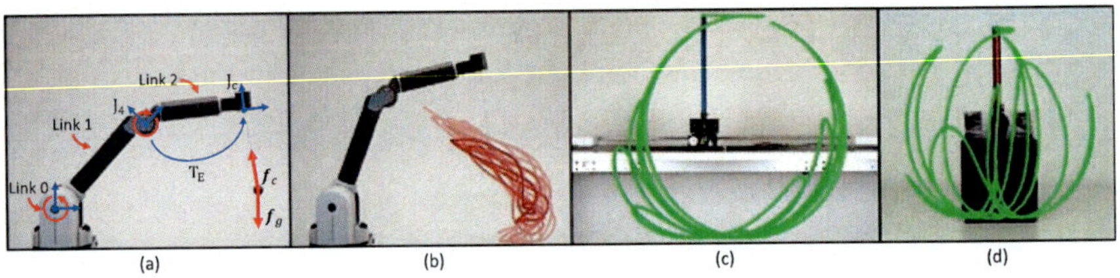

Fig. 2.1 **a** The identified dynamics (red) and kinematic (blue) parameter of the Barrett WAM for the ball in a cup task. **b** Exploration data for the DiffNEA model to infer T_E and the string length. **c** The Quanser cartpole performing an swing-up movement. **d** The Quanser Furuta pendulum performing the swing-up movement. The green trajectory highlights the performed movements of the under-actuated systems

Barrett WAM. The Barrett WAM (Fig. 2.1a, b) consists of four fully actuated degrees of freedom controlled via torque control with 500 Hz. The actuators are back-driveable and consist of cable drives with low gear ratios enabling fast accelerations. The joint angles sensing is on the motor-side. Therefore, any deviation between the motor position and joint position due to the slack of the cables cannot be observed.

Cartpole. The physical cartpole (Fig. 2.1c) is an under-actuated system manufactured by Quanser [66]. The pendulum is passive and the cart is voltage controlled with up to 500 Hz. The linear actuator consists of a plastic cogwheel drive with high stiction.

Furuta Pendulum. The physical Furuta pendulum (Fig. 2.1d) is an under-actuated system manufactured by Quanser [66]. Instead of the linear actuator of the cartpole, the Furuta pendulum has an actuated revolute joint and a passive pendulum. The revolute joint is voltage controlled with up to 500 Hz. The main challenge of this system is the small masses and length scales that requires a sensitive controller.

2.4.1.2 Parametric Dynamics Models

For the evaluation, we compare the performance of the three different instantiations of the DiffNEA model family, a feed-forward neural network (FF-NN) and a LSTM neural network (LSTM). The three variations of DiffNEA are described below

Newton–Euler Algorithm. The Newton–Euler Algorithm (NEA) model assumes knowledge of the kinematic chain and the kinematics parameters $\boldsymbol{\theta}_K$ and only identifies the inertial parameter $\boldsymbol{\theta}_I$. These parameters are identified by linear regression. This model learning approach was concurrently introduced by [4, 18, 36, 53]. Due to the linear regression, this model cannot be augmented with the different actuation models.

No-Kin Differential Newton–Euler Algorithm. The no-Kin DiffNEA model assumes knowledge of the kinematic tree but no knowledge of the kinematics $\boldsymbol{\theta}_K$ or inertial parameters $\boldsymbol{\theta}_I$. The link parameters are identified by minimizing the squared loss of the forward dynamics.

Differential Newton–Euler Algorithm. The DiffNEA model assumes knowledge of the kinematic chain and the kinematics parameters $\boldsymbol{\theta}_K$ and only identifies the inertial parameters $\boldsymbol{\theta}_I$. These parameters are identified by minimizing the squared loss of the forward dynamics.

2.4.1.3 Implementation

All models are implemented using PyTorch [63] and utilize the included automatic differentiation to backpropagate through the equations of motion. For the gradient-based optimization we use ADAM [38] to update the virtual parameters. For the integration of the continuous-time models, we use the Runge–Kutta 4 method (RK4).

2.4.2 Experiment 1—Trajectory Prediction

To evaluate the obtained forward models, we use these models to predict the trajectory of the under-actuated cartpole and Furuta pendulum from an initial state $\boldsymbol{x}_0$ and action sequence $\boldsymbol{u}_{0:T}$. The predicted trajectory is compared to the trajectory of the true model. To evaluate the impact of the dataset, we evaluate the performance on three different datasets with different levels of complexity. The uniform dataset is obtained by sampling joint positions, velocities, and torques from a uniform distribution and computing the acceleration with the true analytic forward dynamics. The simulated trajectory dataset is generated by simulating the ideal system with viscous friction and small state and action noise. The real system dataset is generated on the physical system by applying an energy controller that repeatedly swings up the pendulum and lets the pendulum fall without actuation. In addition, we compare between two initialization strategies, without prior and with prior. Without prior means that the link parameters are initialized randomly. With prior means that the parameters are initialized with the known values given by the manufacturer. This differentiation enables us to evaluate the impact of good initialization for white-box models.

The predicted trajectories and the normalized mean squared error are shown in Fig. 2.2. The overall performance depends on the system as well as the dataset. The numerically sensitive conditioning of the Furuta Pendulum causes all models to be worse on all datasets and model classes. Conversely, the magnitude of the physical parameters of the cartpole makes the identification and long-term prediction simpler. Regarding the datasets, the overall forward prediction performance decreases with dataset complexity. For example, the uniform dataset yields perfect DiffNEA models with very small errors. For the simulated and real trajectory datasets, the prediction error deteriorates. Regarding the different models, no clear best system

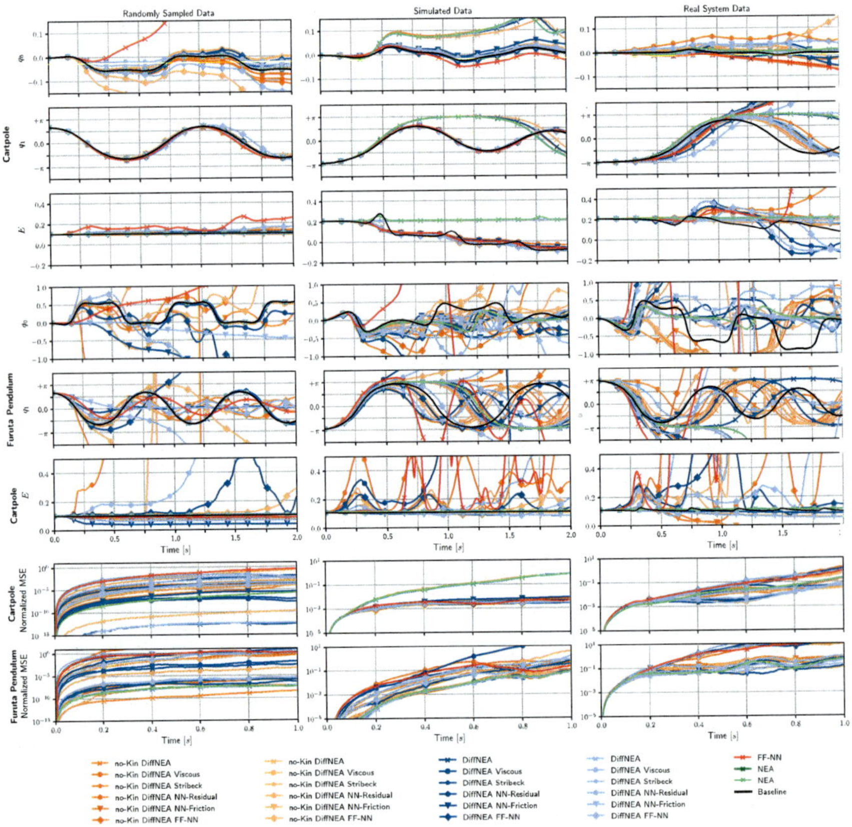

Fig. 2.2 Qualitative model comparison of the 26 different models performing forward roll-outs on the Cartpole and Furuta Pendulum. The difference in the plot colors opacity (for the orange, blue and green traces) highlights whether the model parameters are initialized with the known parameters (low opacity) or random (high opacity). The roll-outs start from the starting state and are computed with 250 Hz sampling frequency and integrated with RK4. The models are trained on three different datasets ranging from uniformly sampled and ideal observations (i.e., simulated data from uniform sampling) to trajectory data of noisy observation from the physical system. The main point of this figure is not to highlight the performance of the individual models but to highlight the distribution of the learned models and the respective failure cases. All models achieve similar performance and fail in many different ways after some system and dataset specific prediction horizon. The model performance and the valid prediction horizon is more affected by the dataset and the specific system characteristics rather than the choice of the model. The identification of the model, performs better for the simulated datasets and the cartpole. The simulated datasets perform better as the true system state is observed and not estimated using noisy sensors. The identification of the cartpole is easier compared to the Furuta pendulum due to the system properties. The mass matrix of the cartpole is well conditioned while the low masses of the Furuta pendulum yield an poorly conditioned mass-matrix. Due to the high sensitivity in parameters, the prediction of the Furuta pendulum have a shorter prediction horizon compared to the cartpole

identification approach is apparent. One can only observe significant differences between the different model classes.

Energy Bounded Versus Energy Unbounded Models. The energy-conserving models (i.e., no-actuator model) perform well when the observed system conserves energy or nearly conserves energy (e.g., Furuta Pendulum). If the systems dissipate a lot of energy, the model prediction degrades. For example, the energy-conserving models do not obtain an accurate prediction for the *simulated* cartpole that contains high viscous friction. For the Furuta pendulum, the friction is negligible, hence the energy-conserving models perform well. The energy-bounded models (i.e., Viscous, Stribeck, and NN-Friction actuator model) are the best performing models of this benchmark. The learned models yield non-diverging trajectories and can model systems with and without friction. The NN-Friction actuator achieves good performance by exploiting its black-box flexibility. Despite its expressiveness, the predicted trajectories do not diverge.

The energy-unbounded models (i.e., FF-NN and NN-Residual actuator model and the black-box model), frequently learn to inject energy into the system even for perfect sensor measurements. For example, the black-box deep network increases the system energy of the cartpole even for the simulated uniformly sampled data. For the more challenging trajectory datasets, all energy-unbounded models predict trajectories that increase in energy during simulation without actuation. Many of these models also generate divergent trajectories during simulation, which leave the training domain.

Learned Kinematics Versus Known Kinematics. One interesting observation is that the DiffNEA model with *unknown* kinematics (no-Kin DiffNEA) performs comparable to the DiffNEA model with *known* kinematics (DiffNEA, NEA), demonstrating the kinematics and dynamics can be learned jointly. However, as the reinforcement learning experiment shows, this observation does not apply to more complex systems. For the ball in a cup experiment, the kinematic structure must be incorporated to obtain a good model of the task. Otherwise one can only identify an accurate model of the WAM but not the cup and string dynamics.

Model Initialization. In the simulation experiments, no clear difference between models with and without prior initialization is visible. Evaluating the identified physical parameters also yields no clear improvement of the initialization with prior, e.g., even for unreasonably large physical parameters, the identified parameter with a prior was not necessarily smaller. Therefore, we conclude that the initialization with the best-known parameters does not necessarily improve model performance.

Conclusion. The experiments show that the DiffNEA model can used for obtaining the system dynamics of systems with nonlinear friction including stiction. The different friction models perform comparably the same in terms of the prediction horizon. This horizon mostly depends on the system characteristics and dataset. For example, the horizon is shorter for the sensitive Furuta pendulum and longer for the cartpole. The horizon is longer for the simulated dataset with uniform state distribution compared to the trajectory datasets. The differences between the models are not significant. The main difference between these models is the behavior beyond this prediction horizon. The energy unbounded models frequently yield diverging trajectories while the energy bounded models do not diverge from the training domain. Therefore, we prefer the NN-friction model. This actuator model combines high model capacity with little assumptions on the friction model and yields non-diverging trajectories.

2.4.3 Experiment 2—Offline Reinforcement Learning

To solve the ball in cup (BiC) task, we use model-based offline reinforcement learning. In this setting, one learns a model from a fixed dataset of arbitrary experience and uses this model to learn the optimal policy [41, 45]. Hence, the agent is bound to a dataset and cannot explore the environment. More specifically, we use episodic relative entropy policy search (eREPS) [64] with an additional KL-divergence constraint on the maximum likelihood policy update [65] and parameter exploration [16] to optimize a distribution over trajectories described using a Probabilistic Movement Primitive (ProMP) [60, 61].

For the manipulator identification the robot executes a 40s high-acceleration sinusoidal joint trajectory. For the string model identification, the robot executes a 40s slow cosine joint trajectories to induce ball oscillation without contact with the manipulator (Fig. 2.1b). The ball trajectories are averaged over five trajectories to reduce the variance of the measurement noise. The training data does not contain swing-up motions and, hence the model must extrapolate to achieve the accurate simulation of the swing-up. The total dataset used for offline RL contains only 4 min of data. To simplify the task for the deep networks, the training data consists of the original training data plus all data generated by the DiffNEA model. Therefore, the network training data contains successful BiC tasks.

The dense episodic reward is inspired by the potential of an electric dipole and augmented with regularizing penalties for joint positions and velocities. The complete reward is defined as

$$R = \exp\left(\frac{1}{2}\max_t \psi_t + \frac{1}{2}\psi_N\right) - \frac{1}{N}\sum_{i=0}^{N} \lambda_q \|\boldsymbol{q}_i - \boldsymbol{q}_0\|_2^2 + \lambda_{\dot{q}} \|\dot{\boldsymbol{q}}_i\|_2^2,$$

with $\psi_t = \Delta_t^\mathsf{T} \hat{\boldsymbol{m}}(\boldsymbol{q}_t) / \left(\Delta_t^\mathsf{T}\Delta_t + \epsilon\right)$ and the normal vector of the end-effector frame $\hat{\boldsymbol{m}}$ which depends on joint configuration $\boldsymbol{q}_t$. For the DiffNEA model, the *predicted* end-effector frame is used during policy optimization. Therefore, the policy is optimized using the reward computed in the approximated model. The black-box models uses the true reward, rather than the reward bootstrapped from the learned model.

For the DiffNEA models, the robot manipulator is modeled as a rigid-body chain using reduced coordinates. The ball is modeled via a constrained particle simulation with an inequality constraint. Both models are interfaced via the task space movement of the robot after the last joint. The manipulator model predicts the task-space movement after the last joint. The string model transforms this movement to the end-effector frame via $\boldsymbol{T}_E$ (Fig. 2.1a), computes the constraint force $\boldsymbol{f}_c$ and the ball acceleration $\ddot{\boldsymbol{x}}_B$. Mathematically this model is described by

$$\ddot{\boldsymbol{x}}_B = \frac{1}{m_B}\left(\boldsymbol{f}_g + \boldsymbol{f}_c\right), \tag{2.12}$$

$$g(\boldsymbol{x}; \boldsymbol{\theta}_S) = \sigma\left(\|\boldsymbol{x}_B - \boldsymbol{T}_E\, \boldsymbol{x}_{J_4}\|_2^2 - r^2\right) = 0, \tag{2.13}$$

where $\boldsymbol{x}_B$ is the ball position, $\boldsymbol{x}_{J_4}$ the position of the last joint and r the string length. In the following we will abbreviate $\boldsymbol{x}_B - \boldsymbol{T}_E\, \boldsymbol{x}_{J_4} = \Delta$ and the cup position by $\boldsymbol{T}_E\, \boldsymbol{x}_{J_4} = \boldsymbol{x}_C$. The constraint force can be computed analytically with the principle of virtual work and is described by

$$\boldsymbol{f}_c(\boldsymbol{\theta}_S) = -m_B\, \sigma'(z)\, \frac{\Delta^\intercal \boldsymbol{g} - \Delta^\intercal \ddot{\mathbf{x}}_C + \dot{\Delta}^\intercal \dot{\Delta}}{\Delta^\intercal \Delta + \delta}, \tag{2.14}$$

with $z = \|\Delta\|_2 - r$, and the gravitational vector $\boldsymbol{g}$. When simulating the system, we set $\ddot{g} = -\boldsymbol{K}_p g - \boldsymbol{K}_d \dot{g} \leq 0$ to avoid constraint violations and add friction to the ball for numerical stability. This closed form constraint force is differentiable and hence one does not need to incorporate any special differentiable simulation variants.

For the black-box models, a feedforward network (FF-NN) and a long short-term memory network (LSTM) [32] is used. The networks model only the string dynamics and receive the task space movement of the last joint and the ball movement as input and predict the ball acceleration, i.e., $\ddot{\boldsymbol{x}}_B = f(\boldsymbol{x}_{J_4}, \dot{\boldsymbol{x}}_{J_4}, \ddot{\boldsymbol{x}}_{J_4}, \boldsymbol{x}_B, \dot{\boldsymbol{x}}_B)$.

For this experiment, we only use the DiffNEA model with known kinematics of the Barrett WAM up to the last joint. Only the end-effector transformation from the last joint to the cup end-effector is identified. Identifying the complete kinematic tree and the dynamics simultaneously, as done in No-Kin DiffNEA in the trajectory prediction experiment, did not yield accurate models as the joint optimization has too much ambiguity.

Simulation Results. The simulation experiments test the models with idealized observations from MuJoCo [78] and enable a quantitative comparison across many seeds. For each model representation, 15 different learned models are evaluated with 150 seeds for the model-free RL. The average statistics of the best ten reinforcement learning seeds are summarized in Table 2.1 and the expected versus obtained reward is shown in Fig. 2.3.

The DiffNEA model can learn the BiC swing-up for every tested model. The transferability to the MuJoCo simulator depends on the specific seed, as the problem contains many different local solutions and only some solutions are robust to slight model variations. The MuJoCo simulator is different from the DiffNEA model as MuJoCo simulates the string as a chain of multiple small rigid bodies. The performance of the identified DiffNEA is comparable to the performance of the DiffNEA model with the true simulator parameters.

Table 2.1 Offline reinforcement learning results for the ball in a cup task, across both simulation and the physical system. Length refers to the string length in centimeters. Repeatability is reported for the best performing reinforcement learning seed. The repeatability of the simulated system is not stated as the simulator is deterministic

Model	Simulation			Physical system			
	Length (cm)	Reward	Transfer (%)	Length (cm)	Reward	Transfer (%)	Repeat (%)
LSTM	40	0.92 ± 0.37	0	40	0.91 ± 0.56	0	0
FF-NN	40	0.86 ± 0.35	0	40	1.46 ± 0.78	0	0
Nominal	40	2.45 ± 1.15	**64**	40	1.41 ± 0.45	0	0
DiffNEA	40	$\mathbf{2.73 \pm 1.64}$	52	40	$\mathbf{1.77 \pm 0.74}$	**60**	90
				35	1.58 ± 0.15	30	70
				45	1.74 ± 0.71	**60**	**100**

The FF-NN and LSTM black-box models do not learn a single successful movement that transfers to the physical system despite using ten different models, 150 different seeds, additional data that includes swing-ups, and observes the real instead of the imagined reward. These learned models cannot stabilize the ball beneath the cup. The ball immediately diverges to a physically unfeasible region. The attached videos compare the real (red) versus imagined (yellow) ball trajectories. Within the impossible region, the policy optimizer exploits the random dynamics where the ball teleports into the cup. Therefore, the policy optimizers converge to random movements.

Real-Robot Results. On the physical Barrett WAM we evaluate 50 seeds per model. A selection of the trials using the identified DiffNEA model is shown in Fig. 2.4. The average statistics of the best ten seeds are summarized in Table 2.1.

The DiffNEA model is capable of solving BiC using offline model-based RL for all three string lengths. This approach obtains very different solutions that transfer to the physical system. Some solutions contain multiple pre-swings which show the quality of the model for long-planning horizons. The best movements achieve the task repeatedly. Solutions that do not transfer to the system, perform feasible movements where the ball bounces off the cup rim. The nominal DiffNEA model with the measured arm and string parameters does not achieve a successful swing-up. The ball always overshoots and bounces off the robot arm for this model.

Similar to the simulation experiments, none of the tested black-box models achieve the BiC swing-up despite using more data and the true rewards during planning. Especially the FF-NN model converges to random policies, which result in ball movements that do not even closely resemble a potential swing-up. The convergence to very different movements shows that the models contain multiple shortcuts capable of teleporting the imagined ball into the cup.

Conclusion. The offline RL experiment shows that the DiffNEA model excels when generalization is required and diverging from the training domain yields spurious

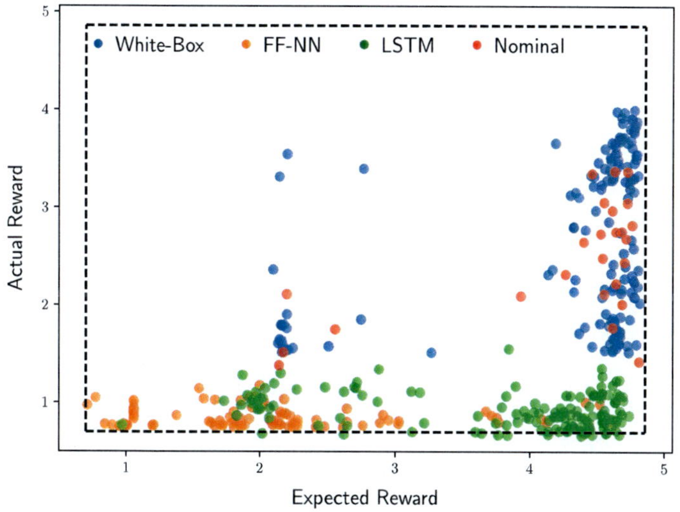

Fig. 2.3 Comparison of the expected reward and the actual reward on the MuJoCo simulator for the LSTM, the feed-forward neural network (FF-NN) as well as the nominal and identified DiffNEA model. The identified and nominal model achieves a comparable performance and solves the BiC swing-up for multiple seeds. Neither the LSTM nor the FF-NN achieve a single successful swing-up despite being repeated with 50 different seeds and using all the data generated by the DiffNEA models

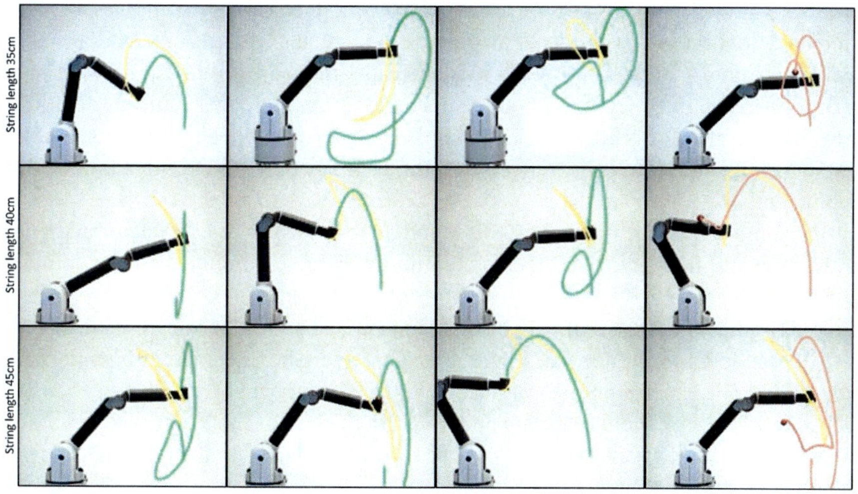

Fig. 2.4 Three different successful swing-ups for the three different string lengths using the DiffNEA model with eREPS for offline model-based reinforcement learning. This approach can learn different swing-ups from just 4 min of data, while all tested black-box models fail at the task. The different solutions are learned using different seeds. The unsuccessful trials of the DiffNEA model nearly solve the BiC tasks but the ball bounces off the cup or arm

solutions. The DiffNEA models achieve the task while black-box models fail. Instead of learning a solution, the model exploitation of black-box models causes the policy optimization to converge to a random movement. The identified DiffNEA models are not perfect. Hence, not all solutions obtained by the RL agent achieve the task when transferred. The transfer rate could be improved by incorporating robustness within the RL agent, e.g., domain randomization or optimizing the worst-case reward given an adversary controlling the dynamics.

2.5 Conclusion

The Differentiable Newton–Euler Algorithm (DiffNEA) learns physically consistent parameters for dissipative mechanical systems with holonomic and non-holonomic constraints. For this approach, we combine differentiable simulation, gradient-based optimization, and virtual parameters to infer physically plausible system parameters of the rigid bodies, the constraints, and the friction models. Therefore, this approach generalizes existing white-box models to a wider class of dynamical systems.

The extensive experiments showed that this model learning technique can learn dynamics models of physical systems with friction and non-holonomic constraints. The learned models can be used for trajectory prediction and reinforcement learning. The obtained DiffNEA models obtain more accurate dynamics models than standard black-box models such as deep networks. However, the accuracy of these DiffNEA models is not necessarily orders of magnitude better than the black-box models under realistic circumstances. Only when model assumptions are valid, the training data is uniformly sampled from the complete state domain and ideal joint observations are obtained, the DiffNEA model identifies near-perfect models. Already when the training data is obtained using trajectories, the accuracy of DiffNEA models degrades significantly.

The main advantage of the DiffNEA models is the worst-case behavior and generalization. While black-box models commonly diverge from the training domain, DiffNEA models always yield physically consistent predictions. The long-term predictions of these models might not be highly accurate but cannot be exploited. Therefore, these models excel for RL, where these characteristics are especially important. The offline RL experiment shows that the DiffNEA model solves the ball in a cup task. The black-box models are not able to solve the task as the agent easily exploits the model and obtains a spurious solution. When transferred to the physical system, the robot performs a random trajectory.

Future Work. To improve system identification techniques for white-box models, one needs to address the ambiguity of the physical parameters. The current approaches that minimize the 1-step error can yield very different parameter configurations as the physical parameters are over parametrized. For small parameters like friction, the 1-step loss is especially problematic as the impact of friction on the 1-step error is often negligible. However, small differences in the friction parameter lead to large errors for the long-term predictions. Therefore, an interesting future research direction is to optimize the system parameters on multiple timescales. Simply using the multi-step loss is not sufficient as backprop through time cannot be applied for very long horizons. Therefore, an interesting research direction for system identification is to leverage the advances of training generative models. For example, one could use a similar approach as generative adversarial networks [25] to identify parameters that yield accurate long-term predictions.

References

1. Albu-Schäffer, A.: Regelung von Robotern mit elastischen Gelenken am Beispiel der DLR-Leichtbauarme. Ph.D. thesis, Technische Universität München (2002)
2. Allevato, A., Short, E.S., Pryor, M., Thomaz, A.: Tunenet: One-shot residual tuning for system identification and sim-to-real robot task transfer. In: Conference on Robot Learning (CoRL) (2020)
3. Åström, K.J., Eykhoff, P.: System identification–a survey. Automatica (1971)
4. Atkeson, C.G., An, C.H., Hollerbach, J.M.: Estimation of inertial parameters of manipulator loads and links. Int. J. Robot. Res. (1986)
5. Atkeson, C.G., Moore, A.W., Schaal, S.: Locally weighted learning for control. Artif. Intell. Rev. (1997)
6. de Avila Belbute-Peres, F., Smith, K., Allen, K., Tenenbaum, J., Kolter, J.Z.: End-to-end differentiable physics for learning and control. In: Advances in Neural Information Processing Systems (2018)
7. Ayusawa, K., Venture, G., Nakamura, Y.: Real-time implementation of physically consistent identification of human body segments. In: IEEE International Conference on Robotics and Automation (ICRA) (2011)
8. Barcelos, L., Oliveira, R., Possas, R., Ott, L., Ramos, F.: Disco: Double likelihood-free inference stochastic control. In: International Conference on Robotics and Automation (ICRA) (2020)
9. Bona, B., Indri, M.: Friction compensation in robotics: an overview. In: 44th IEEE Conference on Decision and Control 2005 and 2005 European Control Conference. CDC-ECC'05. IEEE (2005)
10. Brockett, R.W.: Robotic manipulators and the product of exponentials formula. In: Mathematical Theory of Networks and Systems (1984)
11. Chua, K., Calandra, R., McAllister, R., Levine, S.: Deep reinforcement learning in a handful of trials using probabilistic dynamics models. In: Advances in Neural Information Processing Systems (NeuRIPS) (2018)
12. Coumans, E., Bai, Y.: Pybullet, a python module for physics simulation for games, robotics and machine learning (2016). http://pybullet.org
13. Cranmer, M., Greydanus, S., Hoyer, S., Battaglia, P., Spergel, D., Ho, S.: Lagrangian neural networks (2020). arXiv:2003.04630
14. Degrave, J., Hermans, M., Dambre, J., et al.: A differentiable physics engine for deep learning in robotics. Front. Neurorobot. (2019)
15. Deisenroth, M., Rasmussen, C.E.: Pilco: A model-based and data-efficient approach to policy search. In: Proceedings of the 28th International Conference on machine learning (ICML-11) (2011)
16. Deisenroth, M.P., Neumann, G., Peters, J., et al.: A survey on policy search for robotics. In: Foundations and Trends® in Robotics (2013)
17. Featherstone, R.: Rigid Body Dynamics Algorithms. Springer (2007)
18. Gautier, M.: Identification of robots dynamics. IFAC Proc. Vol. (1986)

19. Gautier, M., Briot, S., Venture, G.: Identification of consistent standard dynamic parameters of industrial robots. In: International Conference on Advanced Intelligent Mechatronics (AIM) (2013)
20. Gautier, M., Venture, G.: Identification of standard dynamic parameters of robots with positive definite inertia matrix. In: International Conference on Intelligent Robots and Systems (IROS) (2013)
21. Gaz, C., Cognetti, M., Oliva, A., Giordano, P.R., De Luca, A.: Dynamic identification of the franka emika panda robot with retrieval of feasible parameters using penalty-based optimization. IEEE Robot. Autom. Lett. (RAL) (2019)
22. Gaz, C., Flacco, F., De Luca, A.: Extracting feasible robot parameters from dynamic coefficients using nonlinear optimization methods. In: IEEE International Conference on Robotics and Automation (ICRA) (2016)
23. Geilinger, M., Hahn, D., Zehnder, J., Bächer, M., Thomaszewski, B., Coros, S.: Add: analytically differentiable dynamics for multi-body systems with frictional contact (2020). arXiv:2007.00987
24. Geist, A.R., Trimpe, S.: Structured learning of rigid-body dynamics: a survey and unified view from a robotics perspective (2020). arXiv:2012.06250
25. Goodfellow, I., Pouget-Abadie, J., Mirza, M., Xu, B., Warde-Farley, D., Ozair, S., Courville, A., Bengio, Y.: Generative adversarial networks. ACM Commun. (2020)
26. Greydanus, S., Dzamba, M., Yosinski, J.: Hamiltonian neural networks. In: Advances in Neural Information Processing Systems (NeuRIPS) (2019)
27. Gupta, J.K., Menda, K., Manchester, Z., Kochenderfer, M.J.: A general framework for structured learning of mechanical systems (2019). arXiv:1902.08705
28. Hafner, D., Lillicrap, T., Ba, J., Norouzi, M.: Dream to control: learning behaviors by latent imagination (2019). arXiv:1912.01603
29. Heiden, E., Macklin, M., Narang, Y., Fox, D., Garg, A., Ramos, F.: Disect: a differentiable simulation engine for autonomous robotic cutting (2021). arXiv:2105.12244
30. Heiden, E., Millard, D., Coumans, E., Sukhatme, G.S.: Augmenting differentiable simulators with neural networks to close the sim2real gap (2020). arXiv:2007.06045
31. Heiden, E., Millard, D., Zhang, H., Sukhatme, G.S.: Interactive differentiable simulation (2019). arXiv:1905.10706
32. Hochreiter, S., Schmidhuber, J.: Long short-term memory. Neural Comput. (1997)
33. Hu, Y., Anderson, L., Li, T.M., Sun, Q., Carr, N., Ragan-Kelley, J., Durand, F.: Difftaichi: differentiable programming for physical simulation (2019). arXiv:1910.00935
34. Hwangbo, J., Lee, J., Dosovitskiy, A., Bellicoso, D., Tsounis, V., Koltun, V., Hutter, M.: Learning agile and dynamic motor skills for legged robots. Sci. Robot. (2019)
35. Jiang, Y., Zhang, T., Ho, D., Bai, Y., Liu, C.K., Levine, S., Tan, J.: Simgan: Hybrid simulator identification for domain adaptation via adversarial reinforcement learning (2021). arXiv:2101.06005
36. Khosla, P.K., Kanade, T.: Parameter identification of robot dynamics. In: Conference on Decision and Control (CDC) (1985)
37. Kim, J.: Lie group formulation of articulated rigid body dynamics. Technical Report, Carnegie Mellon University (2012)
38. Kingma, D.P., Ba, J.: Adam: a method for stochastic optimization (2014). arXiv:1412.6980
39. Kocijan, J., Murray-Smith, R., Rasmussen, C.E., Girard, A.: Gaussian process model based predictive control. In: American Control Conference (ACC). IEEE (2004)
40. Kwon, J., Choi, K., Park, F.C.: Kinodynamic model identification: a unified geometric approach. IEEE Trans. Robot. (T-RO) (2021)
41. Lange, S., Gabel, T., Riedmiller, M.: Batch reinforcement learning. In: Reinforcement Learning: State-of-the-Art. Springer, Berlin, Heidelberg (2012)
42. Lee, S.H., Kim, J., Park, F.C., Kim, M., Bobrow, J.E.: Newton-type algorithms for dynamics-based robot movement optimization. IEEE Trans. Robot. (T-RO) (2005)
43. Lee, T., Park, F.C.: A geometric algorithm for robust multibody inertial parameter identification. IEEE Robot. Autom. Lett. (2018)

44. Lee, T., Wensing, P.M., Park, F.C.: Geometric robot dynamic identification: a convex programming approach. IEEE Trans. Robot. (T-RO) (2019)
45. Levine, S., Kumar, A., Tucker, G., Fu, J.: Offline reinforcement learning: tutorial, review, and perspectives on open problems (2020). arXiv:2005.01643
46. Lutter, M., Hasenclever, L., Byravan, A., Dulac-Arnold, G., Trochim, P., Heess, N., Merel, J., Tassa, Y.: Learning dynamics models for model predictive agents (2021). arXiv:2109.14311
47. Lutter, M., Peters, J.: Deep lagrangian networks for end-to-end learning of energy-based control for under-actuated systems. In: International Conference on Intelligent Robots and Systems (IROS) (2019)
48. Lutter, M., Ritter, C., Peters, J.: Deep lagrangian networks: Using physics as model prior for deep learning. In: International Conference on Learning Representations (ICLR) (2019)
49. Lutter, M., Silberbauer, J., Watson, J., Peters, J.: A differentiable newton euler algorithm for multi-body model learning (2020). arXiv:2010.09802
50. Lutter, M., Silberbauer, J., Watson, J., Peters, J.: Differentiable physics models for real-world offline model-based reinforcement learning. In: International Conference on Robotics and Automation (ICRA) (2021)
51. Mata, V., Benimeli, F., Farhat, N., Valera, A.: Dynamic parameter identification in industrial robots considering physical feasibility. Adv. Robot. (2005)
52. Miller, W.T., Werbos, P.J., Sutton, R.S.: Neural Networks for Control. MIT Press (1995)
53. Mukerjee, A., Ballard, D.: Self-calibration in robot manipulators. In: International Conference on Robotics and Automation (ICRA) (1985)
54. Muratore, F., Gruner, T., Wiese, F., Belousov, B., Gienger, M., Peters, J.: Neural posterior domain randomization. In: Conference on Robot Learning (CoRL) (2021)
55. Nguyen-Tuong, D., Peters, J.: Using model knowledge for learning inverse dynamics. In: International Conference on Robotics and Automation (2010)
56. Nguyen-Tuong, D., Peters, J.: Model learning for robot control: a survey. Cognit. Proc. (2011)
57. Nguyen-Tuong, D., Seeger, M., Peters, J.: Computed torque control with nonparametric regression models. In: American Control Conference (ACC). IEEE (2008)
58. Nguyen-Tuong, D., Seeger, M., Peters, J.: Model learning with local gaussian process regression. Adv. Robot. (2009)
59. Olsson, H., Åström, K.J., De Wit, C.C., Gäfvert, M., Lischinsky, P.: Friction models and friction compensation. Eur. J. Control (1998)
60. Paraschos, A., Daniel, C., Peters, J., Neumann, G.: Using probabilistic movement primitives in robotics. Auton. Robots (2018)
61. Paraschos, A., Daniel, C., Peters, J.R., Neumann, G.: Probabilistic movement primitives. In: Advances in Neural Information Processing Systems (NeurIPS) (2013)
62. Park, F.C., Bobrow, J.E., Ploen, S.R.: A lie group formulation of robot dynamics. Int. J. Robot. Res. (IJRR) (1995)
63. Paszke, A., Gross, S., Massa, F., Lerer, A., Bradbury, J., Chanan, G., Killeen, T., Lin, Z., Gimelshein, N., Antiga, L., Desmaison, A., Kopf, A., Yang, E., DeVito, Z., Raison, M., Tejani, A., Chilamkurthy, S., Steiner, B., Fang, L., Bai, J., Chintala, S.: PyTorch: an imperative style, high-performance deep learning library. In: Advances in Neural Information Processing Systems 32 (2019)
64. Peters, J., Mulling, K., Altun, Y.: Relative entropy policy search. In: Conference on Artificial Intelligence (AAAI) (2010)
65. Ploeger, K., Lutter, M., Peters, J.: High acceleration reinforcement learning for real-world juggling with binary rewards. In: Conference on Robot Learning (Corl) (2020)
66. Quanser: Quanser courseware and resources (2018). https://www.quanser.com/solution/control-systems/
67. Rall, L.B.: Automatic Differentiation: Techniques and Applications. Lecture Notes in Computer Science. Springer (1981)
68. Ramos, F., Possas, R.C., Fox, D.: Bayessim: adaptive domain randomization via probabilistic inference for robotics simulators (2019). arXiv:1906.01728

69. Saemundsson, S., Terenin, A., Hofmann, K., Deisenroth, M.: Variational integrator networks for physically structured embeddings. In: International Conference on Artificial Intelligence and Statistics (Aistats) (2020)
70. Sanchez-Gonzalez, A., Godwin, J., Pfaff, T., Ying, R., Leskovec, J., Battaglia, P.W.: Learning to simulate complex physics with graph networks (2020). arXiv:2002.09405
71. Sanchez-Gonzalez, A., Heess, N., Springenberg, J.T., Merel, J., Riedmiller, M., Hadsell, R., Battaglia, P.: Graph networks as learnable physics engines for inference and control (2018). arXiv:1806.01242
72. Schaal, S., Atkeson, C.G., Vijayakumar, S.: Scalable techniques from nonparametric statistics for real time robot learning. Appl. Intell. (2002)
73. Sousa, C.D., Cortesao, R.: Physical feasibility of robot base inertial parameter identification: a linear matrix inequality approach. Int. J. Robot. Res. (IJRR) (2014)
74. Sousa, C.D., Cortesao, R.: Inertia tensor properties in robot dynamics identification: A linear matrix inequality approach. IEEE/ASME Transactions on Mechatronics (2019)
75. Spong, M.W., Hutchinson, S., Vidyasagar, M.: Robot Modeling and Control. Wiley (2020)
76. Sutanto, G., Wang, A., Lin, Y., Mukadam, M., Sukhatme, G., Rai, A., Meier, F.: Encoding physical constraints in differentiable Newton-Euler algorithm. In: Learning for Dynamics Control (L4DC) (2020)
77. Ting, J.A., Mistry, M., Peters, J., Schaal, S., Nakanishi, J.: A bayesian approach to nonlinear parameter identification for rigid body dynamics. In: Robotics: Science and Systems (2006)
78. Todorov, E., Erez, T., Tassa, Y.: Mujoco: a physics engine for model-based control. In: International Conference on Intelligent Robots and Systems (IROS) (2012)
79. Traversaro, S., Brossette, S., Escande, A., Nori, F.: Identification of fully physical consistent inertial parameters using optimization on manifolds. In: International Conference on Intelligent Robots and Systems (IROS) (2016)
80. Turrisi, G., Capotondi, M., Gaz, C., Modugno, V., Oriolo, G., De Luca, A.: On-line learning for planning and control of underactuated robots with uncertain dynamics. IEEE Robot. Autom. Lett. (RAL) (2021)
81. Wahrburg, A., Bös, J., Listmann, K.D., Dai, F., Matthias, B., Ding, H.: Motor-current-based estimation of cartesian contact forces and torques for robotic manipulators and its application to force control. IEEE Trans. Autom. Sci. Eng. (2018)
82. Wensing, P.M., Kim, S., Slotine, J.J.E.: Linear matrix inequalities for physically consistent inertial parameter identification: a statistical perspective on the mass distribution. IEEE Robot. Autom. Lett. (RAL) (2017)
83. Werling, K., Omens, D., Lee, J., Exarchos, I., Liu, C.K.: Fast and feature-complete differentiable physics for articulated rigid bodies with contact (2021). arXiv:2103.16021
84. Yoshida, K., Khalil, W.: Verification of the positive definiteness of the inertial matrix of manipulators using base inertial parameters. Int. J. Robot. Res. (2000)
85. Zhong, Y.D., Dey, B., Chakraborty, A.: Symplectic ode-net: learning hamiltonian dynamics with control (2019). arXiv:1909.12077

Chapter 3
Combining Physics and Deep Learning for Continuous-Time Dynamics Models

3.1 Introduction

During the last five years, deep learning has shown the potential to fundamentally change the use of learning in robotics. Currently, many robot learning approaches involve a deep network as part of their algorithm. The network either represents a policy that selects the actions, a dynamics model that predicts the next state, or a state estimator that extracts the relevant features from unstructured observations. Initially, many of these approaches were only applicable to simulated environments due to the large amounts of data required to train the networks. When using massive parallelized simulations, these methods achieved astonishing results [42]. By now, these learning algorithms have been improved and start to be applied to real-world systems [1, 37]. On the physical systems, the deep network approaches have not bypassed classical robotics techniques yet, but have shown very promising results achieving comparable results as classical methods.

Within many proposed algorithms deep networks have replaced analytic models and other function approximators due to their simplicity, generic applicability, scalability, high model capacity and widespread availability of GPU's enabling fast training and evaluation. The generic applicability of these black-box models combined with the high model capacity is a curse and blessing. On the one side, this combination enables the learning of arbitrary functions with high fidelity. However, this combination is also susceptible to overfit to the data without retrieving the underlying structure. Furthermore, the black-box nature of standard deep networks prevents including prior knowledge from first-order principles. This limitation is especially problematic for robotics as the overfitting to spurious data can lead to unpredictable behaviors damaging the physical system. The problem is also made unnecessarily harder as all existing knowledge of robotics and mechanics is ignored.

In this chapter, we propose a new approach that combines existing knowledge with deep networks. This combination enables to learn better representations for robotics and retains the advantages of deep networks. To learn physically plausible

M. Lutter, *Inductive Biases in Machine Learning for Robotics and Control*,
Springer Tracts in Advanced Robotics 156,
https://doi.org/10.1007/978-3-031-37832-4_3

continuous-time dynamics models of rigid body systems, we combine Lagrangian mechanics with deep networks. The proposed Deep Lagrangian Networks (DeLaN) use two deep networks to parameterize the kinetic and potential energy [68]. The network parameters are learned by minimizing the squared residual of the Euler–Lagrange differential equation. The resulting dynamics models are guaranteed to evolve as a mechanical system and conserve the system energy. Therefore, these models achieve better long-term predictions and control performance than the standard black-box models. The resulting physics-inspired models share many of the characteristics of analytic models without requiring specific knowledge about the individual system. For example, DeLaN models are interpretable and enable the computation of the gravitational forces, the momentum, and the system energy. Previously, computing this decomposition was only possible using the analytic models with the system parameters. DeLaN also enables the computation of the forward and inverse models with the same parameters. These characteristics are in stark contrast to standard black-box models. Such black-box models only obtain either the forward or the inverse model and cannot compute the different physical quantities as these need to be learned unsupervised. Due to these advantages of physics-inspired dynamics models, many variants have been proposed [21, 33, 35, 86, 106].

3.1.1 Outline

To provide a self-contained overview about physics-inspired deep networks for learning dynamics models, we briefly summarize the related work (Sect. 3.2), prior approaches for learning dynamics models of rigid body systems as well as the basics of Lagrangian and Hamiltonian mechanics (Sect. 3.3.2). Subsequently, we introduce physics-inspired networks derived from Lagrangian and Hamiltonian mechanics as well as the existing variants (Sect. 3.4). Section 3.5 presents the experimental results of applying these models to model-based control and compares the performance to system identification as well as deep network dynamics models. Finally, Sect. 3.6 discusses the experimental results, highlights the limitations of physics-inspired networks, and summarizes the contributions of this chapter.

3.2 Related Work

In the main part of this chapter, we focus on learning continuous-time dynamics models of mechanical systems. However, physics-inspired networks and continuous-time deep networks have been utilized for different applications areas. In this section, we want to briefly summarize the existing work on both topics outside the domain of rigid body systems and their differences.

3.2.1 Physics-Inspired Deep Networks

Incorporating knowledge of physics within deep networks has been approached by introducing conservation laws or symmetries within the network architecture. Both approaches are tightly coupled due to Noether's theorem showing that symmetries induce conservation laws. In the case of conservation laws, these laws can be incorporated by minimizing the residual of the corresponding differential equation to obtain the optimal network parameters. The combination of deep learning and differential equations has been well known for a long time and investigated in more abstract forms [56, 57, 60, 71]. Using this approach, various authors proposed to use the Navier–Stokes equation [17, 80], Schroedinger equation [81], Burgers Equation [47], Hamilton's equation [15, 33, 95, 106] or the Euler–Lagrange equation [21, 35, 68, 78].

Symmetries can be integrated within the network architecture by selecting a non-linear transformation that is either equivariant, i.e., preserves the symmetry, or is invariant to specific transformations of the input. Using this approach, one can derive layers that are translational-, rotational-, scale- and gauge equivariant [7, 18, 19, 99]. These architectures are frequently used for computer vision as image classification is translational and rotational invariant [18, 62, 100]. Up to now, only very few papers have applied this approach to model physical phenomena [5, 99]. A different approach to symmetries was proposed by [48]. To obtain time translation invariance, which is equivalent to conservation of energy, this work optimized time-reversibility. Therefore, the symmetry is not incorporated in the network architecture but the optimization loss.

Besides these generic approaches utilizing symmetries and conservation laws, various authors also proposed specific architectures for individual problems. In this case, the known spatial structure of the problem is embedded within the network architecture. For example, [98] proposed a network architecture for turbulent flow predictions that incorporates multiple spatial and temporal scales. Reference [88] used a graph network to encode the known kinematic structure and the local interactions between two links within the network structure. Similarly, [90] incorporates the local structure of molecules within the network architecture.

3.2.2 Continuous-Time Models and Neural ODEs

The work on neural ordinary differential equation (ODE) by [14] initiated a large surge of research on continuous-time models. The original work on neural ODE proposed a deep network with infinite depth to improve classification and density estimation. While these algorithms were not meant for modeling dynamical systems, the explicit integration step within the neural ODE led to rediscover continuous-time models for dynamical systems. Since then, neural ode's have been frequently mentioned as inspiration to learn continuous-time models [10, 46, 48, 86]. Frequently the

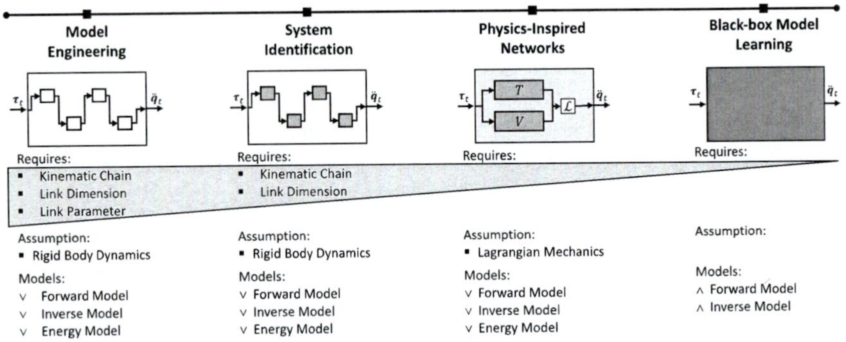

Fig. 3.1 The requirements and assumptions of the different approaches to obtain the dynamics model of mechanical systems. The physics-inspired networks bridge the gap between classical system identification and black-box model learning. While system identification requires knowledge of the kinematic chain, the physics-inspired networks do not require any knowledge of the specific system but obtain comparable characteristics as system identification. Physics-inspired networks also guarantee energy-conserving models and obtain the forward, inverse, and energy model simultaneously

term neural ODE is used interchangeably for a continuous-time model with a deep network. In this work, we will only use the term continuous-time model. One technical difference between the original neural ODE and continuous-time models is that the neural ODE uses a variable time step integrator, most commonly the Dormand–Prince method. The continuous-time models use a fixed time step integrator. For the fixed time step integrator, different authors have used the explicit Euler, the Runge Kutta method, or symplectic integrators. For dynamics models, the fixed time step is convenient as the data is observed at a fixed time step determined by the sampling rate of digital sensors.

3.3 Preliminaries

We want to introduce the standard model learning techniques for dynamical systems and briefly summarize the relevant theory of Lagrangian and Hamiltonian Mechanics.

3.3.1 *Learning Dynamics Models*

Models describing system dynamics, i.e. the coupling of the system input $\boldsymbol{u}$ and system state $\boldsymbol{x}$, are essential for model-based control approaches [50] and model based planning. Depending on the approach, one either relies on the forward or inverse model. For example, inverse dynamics control [105] uses the inverse model to compensate system dynamics, while model-predictive control [12] and optimal control [109] use the forward model to compute the future states given an action sequence. For discrete time models, the forward model f maps from the system state $\boldsymbol{x}_t$ and input $\boldsymbol{u}_t$ to the next state $\boldsymbol{x}_{t+1}$. The inverse model f^{-1} maps from system state and the next state to the system input. Mathematically this is described by

$$f(\boldsymbol{x}_t, \boldsymbol{u}_t;\ \boldsymbol{\theta}) = \boldsymbol{x}_{t+1}, \qquad f^{-1}(\boldsymbol{x}_t, \boldsymbol{x}_{t+1};\ \boldsymbol{\theta}) = \boldsymbol{u}_t, \tag{3.1}$$

with the model parameters $\boldsymbol{\theta}$. In the continuous time setting the next state $\boldsymbol{x}_{t+1}$ is replaced with the change of the state $\dot{\boldsymbol{x}}$, i.e.,

$$f(\boldsymbol{x}_t, \boldsymbol{u}_t;\ \boldsymbol{\theta}) = \dot{\boldsymbol{x}}_t, \qquad f^{-1}(\boldsymbol{x}_t, \dot{\boldsymbol{x}}_t;\ \boldsymbol{\theta}) = \boldsymbol{u}_t. \tag{3.2}$$

The continuous-time system can be combined with an integrator, e.g., explicit Euler, Runge–Kutta method, or symplectic integrators, to predict the next state instead of the change of the system state. Therefore, the continuous-time model is independent of the time discretization. Depending on the chosen representation, the transfer function f and parameters $\boldsymbol{\theta}$ are obtained using different approaches. In the following, we will differentiate the different approaches, (1) model engineering, (2) data-driven system identification, and (3) black-box model learning (Fig. 3.1). In Sect. 3.4, we will extend these existing categories to physics-inspired models.

3.3.1.1 Model Engineering

The most classical approach is model engineering, which is predominantly used within the industry. In this case, the transfer function f is the equations of motion and the model parameters are the physical parameters of the robot consisting of the masses, center of gravity, length, and inertia. The equations of motion have to be manually derived for each system. Frequently, one assumes perfect rigid bodies connected by ideal joints and uses Newtonian, Lagrangian or Hamiltonian mechanics and the known structure of the systems to derive the equations. The model parameters can be either inferred using the CAD software or measured by disassembling the individual system. The latter is more precise as it incorporates the deviations due to the manufacturing process [2]. Furthermore, the parameters are identical for the forward and inverse model. Therefore, this approach yields the forward and inverse model simultaneously. To summarize, this approach can yield very precise forward and inverse models for rigid body systems but is labor-intensive as the parameters need to be manually inferred.

3.3.1.2 Data-Driven System Identification

Similar to model engineering, data-driven system identification uses the analytic equations of motions as the transfer function. However, the model parameters are learned from observed data rather than measured. Therefore, the equations of motions need to be manually derived but the model parameters are learned. In 1985 four different groups showed concurrently that the dynamics parameters can be obtained by linear regression using hand-designed features for rigid body kinematic chains [6, 28, 54, 73]. This approach is commonly referenced as the standard system identification technique for robot manipulators by the textbooks [91]. However, this approach cannot guarantee physically plausible parameters as the dynamics parameters have additional constraints. For example, this approach can yield negative masses, an inertia matrix that is not positive definite, or violate the parallel axis theorem [94]. The disadvantages of this approach are that one can only infer linear combinations of the dynamics parameters, cannot apply it to close-loop kinematics [91] and can only be applied inverse dynamics. The inverse dynamic formulation is problematic as the inverse dynamics do not necessarily have a unique solution due to friction [82]. To overcome these shortcomings, [94] proposed a projection-based approach while many others [29, 59, 69, 70, 93, 96, 102] used virtual parametrizations that guarantee physical plausible parameters. For the latter, the optimization does not simplify to linear regression but can be solved using gradient descent. To summarize, this approach only requires the equations of motions analytically and can learn the dynamical parameters from data. Therefore, this approach is not as labor-intensive as model engineering but one needs to ensure to collect 'good' data for the learning.

3.3.1.3 Black-Box Model Learning

While the previous approaches required knowledge about the individual kinematic chain to derive the equations of motion, the black-box approaches do not require any knowledge of the system. These approaches use any black-box function approximator as a transfer function and optimize the model parameters to fit the observed data. For example, the existing literature used Local Linear Models [41, 89], Gaussian Mixture Models [11, 53], Gaussian Processes [13, 55, 74, 76, 83, 84], Support Vector Machines [16, 25], feedforward- [51, 59, 63, 88] or recurrent neural networks [36, 38, 39, 85] to learn the dynamics model. The black-box models obtain either the forward or inverse model and the learned model is only valid on the training data distribution. The previous methods based on the analytic equations of motions obtained both models simultaneously and generalize beyond the data distribution as the learned physical parameters are globally valid. However, the black-box models do not require assumptions about the systems and can learn systems including contacts. These previous approaches relied on assuming rigid body dynamics and could only learn the system dynamics of articulated bodies using reduced coordinates without contacts. Therefore, black-box models can be more accurate for real-world systems where the underlying assumption is not valid but is limited to the training domain and rarely extrapolate.

3.3.2 Lagrangian Mechanics

One approach to derive equations of motion is Lagrangian Mechanics. In the following, we summarize this approach as we will use it in Sect. 3.4 to propose a physics-inspired network for learning dynamics models. More specifically we use the Euler–Lagrange formulation with non-conservative forces and generalized coordinates. For more information and the formulation using Cartesian coordinates please refer to the textbooks [24, 32, 105]. Generalized coordinates $\boldsymbol{q}$ are coordinates that uniquely define the system configuration without constraints. These coordinates are often called reduced coordinates. For articulated bodies, the system state can be expressed as $\boldsymbol{x} = [\boldsymbol{q}, \dot{\boldsymbol{q}}]$. The Lagrangian mechanic's formalism defines the Lagrangian $\mathcal{L}$ as a function of generalized coordinates $\boldsymbol{q}$ describing the complete dynamics of a given system. The Lagrangian is not unique and every $\mathcal{L}$ which yields the correct equations of motion is valid. The Lagrangian is generally chosen to be

$$\mathcal{L}(\boldsymbol{q}, \dot{\boldsymbol{q}}) = T(\boldsymbol{q}, \dot{\boldsymbol{q}}) - V(\boldsymbol{q}) = \frac{1}{2}\dot{\boldsymbol{q}}^\top \mathbf{H}(\boldsymbol{q})\dot{\boldsymbol{q}} - V(\boldsymbol{q}), \tag{3.3}$$

with the kinetic energy T, the potential energy V and the mass matrix $\mathbf{H}(\boldsymbol{q})$. The kinetic energy T is quadratic for any choice of generalized coordinates and any non-relativistic system. The mass matrix is the symmetric and positive definite [105]. The positive definiteness ensures that all non-zero velocities lead to positive kinetic energy. Applying the calculus of variations yields the Euler–Lagrange equation with non-conservative forces described by

$$\frac{d}{dt}\frac{\partial \mathcal{L}(\boldsymbol{q}, \dot{\boldsymbol{q}})}{\partial \dot{\boldsymbol{q}}} - \frac{\partial \mathcal{L}(\boldsymbol{q}, \dot{\boldsymbol{q}})}{\partial \boldsymbol{q}} = \boldsymbol{\tau}, \tag{3.4}$$

$$\frac{\partial^2 \mathcal{L}(\boldsymbol{q}, \dot{\boldsymbol{q}})}{\partial^2 \dot{\boldsymbol{q}}}\ddot{\boldsymbol{q}} + \frac{\partial \mathcal{L}(\boldsymbol{q}, \dot{\boldsymbol{q}})}{\partial \boldsymbol{q}\partial \dot{\boldsymbol{q}}}\dot{\boldsymbol{q}} - \frac{\partial \mathcal{L}(\boldsymbol{q}, \dot{\boldsymbol{q}})}{\partial \boldsymbol{q}} = \boldsymbol{\tau}, \tag{3.5}$$

where $\boldsymbol{\tau}$ are generalized forces frequently corresponding to the system input $\boldsymbol{u}$. Substituting $\mathcal{L}$ with the kinetic and potential energy into Eq. 3.4 yields the second order ODE described by

$$\mathbf{H}(\boldsymbol{q})\ddot{\boldsymbol{q}} + \underbrace{\dot{\mathbf{H}}(\boldsymbol{q})\dot{\boldsymbol{q}} - \frac{1}{2}\left(\dot{\boldsymbol{q}}^\top \frac{\partial \mathbf{H}(\boldsymbol{q})}{\partial \boldsymbol{q}}\dot{\boldsymbol{q}}\right)^\top}_{=\mathbf{c}(\boldsymbol{q},\dot{\boldsymbol{q}})} + \underbrace{\frac{\partial V(\boldsymbol{q})}{\partial \boldsymbol{q}}}_{=\mathbf{g}(\boldsymbol{q})} = \boldsymbol{\tau}, \tag{3.6}$$

where $\mathbf{c}$ describes the forces generated by the Centripetal and Coriolis forces and $\boldsymbol{g}$ the gravitational forces [24]. Most robotics textbooks abbreviate this equation as $\boldsymbol{H}(\boldsymbol{q})\ddot{\boldsymbol{q}} + \boldsymbol{c}(\boldsymbol{q}, \dot{\boldsymbol{q}}) + \boldsymbol{g}(\boldsymbol{q}) = \boldsymbol{\tau}$. Using this ODE, any multi-particle mechanical system with holonomic constraints can be described. Various authors used this ODE to manually derive the equations of motion for coupled pendulums [32], robotic manipulators with flexible joints [9, 92], parallel robots [30, 64, 72] or legged robots [31, 44].

3.3.3 Hamiltonian Mechanics

A different approach to deriving the equations of motions is Hamiltonian mechanics. In this case, the system dynamics are described using the state $\boldsymbol{x} = [\boldsymbol{q}, \boldsymbol{p}]$ with generalized momentum $\boldsymbol{p}$ instead of the generalized velocity $\dot{\boldsymbol{q}}$ and the Hamiltonian $\mathcal{H}$ instead of the Lagrangian. The generalized momentum can be expressed using the Lagrangian and is described by $\boldsymbol{p} = \partial\mathcal{L}/\partial\dot{\boldsymbol{q}}$ [27]. Given the parametrization of the Lagrangian (Eq. 3.3), this definition is equivalent to $\boldsymbol{p} = \boldsymbol{H}(\boldsymbol{q})\,\dot{\boldsymbol{q}}$. The Hamiltonian describes the complete energy of the system and is defined as

$$\mathcal{H}(\boldsymbol{q}, \boldsymbol{p}) = T(\boldsymbol{q}, \boldsymbol{p}) + V(\boldsymbol{q}) = \frac{1}{2}\boldsymbol{p}^{\top}\boldsymbol{H}(\boldsymbol{q})^{-1}\boldsymbol{p} + V(\boldsymbol{q}). \tag{3.7}$$

The Hamiltonian can be computed by applying the Legendre transformation to the Lagrangian which is described by

$$\mathcal{H}(\boldsymbol{q}, \dot{\boldsymbol{q}}) = \dot{\boldsymbol{q}}^{\top}\frac{\partial\mathcal{L}(\boldsymbol{q}, \dot{\boldsymbol{q}})}{\partial\dot{\boldsymbol{q}}} - \mathcal{L}(\boldsymbol{q}, \dot{\boldsymbol{q}}). \tag{3.8}$$

Using the generalized momentum $\boldsymbol{p}$ and the generalized coordinate $\boldsymbol{q}$, the Euler–Lagrange equation can be rewritten to yield Hamilton's equations with control [32]. Hamilton's equations is described by

$$\dot{\boldsymbol{q}} = \frac{\partial\mathcal{H}(\boldsymbol{q}, \boldsymbol{p})}{\partial\boldsymbol{p}}, \qquad \dot{\boldsymbol{p}} = -\frac{\partial\mathcal{H}(\boldsymbol{q}, \boldsymbol{p})}{\partial\boldsymbol{q}} + \boldsymbol{\tau}. \tag{3.9}$$

The Euler–Lagrange equation (Eq. 3.6) can be easily derived from Hamilton's equation by substituting Eq. 3.7 into Eq. 3.9 and using the definition of the generalized momentum, i.e.,

$$\begin{aligned}
\dot{\boldsymbol{p}} &= -\frac{\partial\mathcal{H}(\boldsymbol{q}, \boldsymbol{p})}{\partial\boldsymbol{q}} + \boldsymbol{\tau}, \\
\frac{d}{dt}\left[\boldsymbol{H}(\boldsymbol{q})\,\dot{\boldsymbol{q}}\right] &= \frac{1}{2}\left(\boldsymbol{p}^{\top}\boldsymbol{H}^{-1}\frac{\partial\boldsymbol{H}}{\partial\boldsymbol{q}}\boldsymbol{H}^{-1}\boldsymbol{p}\right)^{\top} - \frac{\partial V}{\partial\boldsymbol{q}} + \boldsymbol{\tau}, \\
\boldsymbol{H}\ddot{\boldsymbol{q}} + \dot{\boldsymbol{H}}\dot{\boldsymbol{q}} &= \frac{1}{2}\left(\dot{\boldsymbol{q}}^{\top}\frac{\partial\boldsymbol{H}}{\partial\boldsymbol{q}}\dot{\boldsymbol{q}}\right)^{\top} - \frac{\partial V}{\partial\boldsymbol{q}} + \boldsymbol{\tau}, \\
\boldsymbol{H}\ddot{\boldsymbol{q}} + \dot{\boldsymbol{H}}\dot{\boldsymbol{q}} - \frac{1}{2}\left(\dot{\boldsymbol{q}}^{\top}\frac{\partial\boldsymbol{H}}{\partial\boldsymbol{q}}\dot{\boldsymbol{q}}\right)^{\top} + \frac{\partial V}{\partial\boldsymbol{q}} &= \boldsymbol{\tau}.
\end{aligned}$$

Many textbooks omit the generalized forces within Hamilton's equation but adding these generalized forces is straightforward as shown in the previous derivation.

3.4 Physics-Inspired Deep Networks

A different approach to black-box model learning is to combine black-box models with physics to guarantee a physically plausible dynamics model. One combination is to use deep networks to represent the system energy and use the resulting Lagrangian to derive the equations of motion using the Euler–Lagrange differential equation. This approach was initially proposed by [68] with the presentation of Deep Lagrangian Networks (DeLaN). Since then, many papers exploring variations of these approaches have been proposed including approaches that use Hamiltonian mechanics instead of Lagrangian mechanics [21, 33, 35, 46, 86, 87, 106–108].

All of these models have in common, that the learned dynamics models conserve energy when the non-conservative forces can be modeled properly and the generalized coordinates are observed. Therefore, the learned model is guaranteed to adhere to one of the fundamental concepts of physics. This property is beneficial as it has been shown within prior research that naive deep network dynamics models frequently increase or decrease the system energy even when the energy should be conserved [33, 46, 86, 106].

In the following, we present DeLaN (Sect. 3.4.1) and the combination of Hamiltonian mechanics and deep networks in Sect. 3.4.2. Afterwards, Sect. 3.4.3 describes all the proposed extensions of DeLaN and Hamiltonian Neural Networks (HNN). Therefore, this section provides the big picture of existing physics-inspired deep networks for learning dynamics models. The flow charts of the variants are shown in Fig. 3.2.

3.4.1 *Deep Lagrangian Networks (DeLaN)*

DeLaN is one instantiation of these physics-inspired deep networks. DeLaN parametrizes the mass matrix $\boldsymbol{H}$ and the potential energy V as two separate deep networks. Therefore, the approximate Lagrangian $\mathcal{L}$ described by

$$\mathcal{L}(\boldsymbol{q}, \dot{\boldsymbol{q}};\ \boldsymbol{\psi}, \boldsymbol{\phi}) = \frac{1}{2}\dot{\boldsymbol{q}}^\top \boldsymbol{H}(\boldsymbol{q};\ \boldsymbol{\psi})\,\dot{\boldsymbol{q}} - V(\boldsymbol{q};\ \boldsymbol{\phi}). \tag{3.10}$$

Using this parametrization the forward and inverse model can be derived. The forward model $\ddot{\boldsymbol{q}} = f(\boldsymbol{q}, \dot{\boldsymbol{q}}, \boldsymbol{\tau};\ \boldsymbol{\psi}, \boldsymbol{\phi})$ is described by

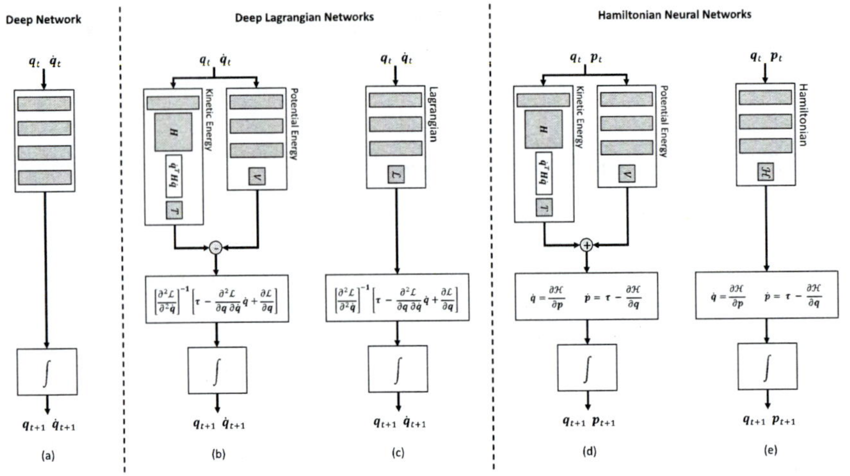

Fig. 3.2 The flowcharts of a continuous-time forward model using a deep network and the physics-inspired networks forward models. **a** Standard deep model learning approach, where a network is used to directly predict the change in position and velocity. **b–c** Deep Lagrangian Networks which use deep networks to predict the Lagrangian $\mathcal{L}$ of the dynamical system and compute the change in position and velocity using the Euler–Lagrange differential equation. **b** Shows the structured Lagrangian approach, where two networks predict the kinetic T and potential energy V and computes the Lagrangian is analytically using $\mathcal{L} = T - V$. **c** Shows the black-box Lagrangian approach where a network directly predicts $\mathcal{L}$. **d–e** Hamiltonian Neural networks which use deep networks to predict the Hamiltonian $\mathcal{H}$ and computes the change in position and impulse via Hamilton's equation. Similar to the Lagrangian variants, **d** shows the structured Hamiltonian and **e** the black-box Hamiltonian. The structured Hamiltonian computes the Hamiltonian via $\mathcal{H} = T(\boldsymbol{q}, \dot{\boldsymbol{q}}) - V(\boldsymbol{q})$, where the kinetic and potential energy is predicted using two networks. The black-box Hamiltonian uses a single network to directly predict $\mathcal{H}$

$$
\begin{aligned}
\ddot{\boldsymbol{q}} &= \left[\frac{\partial^2 \mathcal{L}(\boldsymbol{q}, \dot{\boldsymbol{q}})}{\partial^2 \dot{\boldsymbol{q}}}\right]^{-1} \left[\boldsymbol{\tau} - \frac{\partial \mathcal{L}(\boldsymbol{q}, \dot{\boldsymbol{q}})}{\partial \boldsymbol{q} \partial \dot{\boldsymbol{q}}} \dot{\boldsymbol{q}} + \frac{\partial \mathcal{L}(\boldsymbol{q}, \dot{\boldsymbol{q}})}{\partial \boldsymbol{q}}\right], \\
&= \quad \boldsymbol{H}^{-1} \quad \left[\boldsymbol{\tau} - \dot{\boldsymbol{H}} \dot{\boldsymbol{q}} + \frac{1}{2}\left(\dot{\boldsymbol{q}}^{\top} \frac{\partial \mathbf{H}}{\partial \boldsymbol{q}} \dot{\boldsymbol{q}}\right)^{\top} - \frac{\partial V}{\partial \boldsymbol{q}}\right].
\end{aligned}
$$

The inverse model $\boldsymbol{\tau} = f^{-1}(\boldsymbol{q}, \dot{\boldsymbol{q}}, \ddot{\boldsymbol{q}};\ \boldsymbol{\psi}, \boldsymbol{\phi})$ is described by

$$
\begin{aligned}
\boldsymbol{\tau} &= \frac{\partial^2 \mathcal{L}(\boldsymbol{q}, \dot{\boldsymbol{q}})}{\partial^2 \dot{\boldsymbol{q}}} \ddot{\boldsymbol{q}} + \frac{\partial \mathcal{L}(\boldsymbol{q}, \dot{\boldsymbol{q}})}{\partial \boldsymbol{q} \partial \dot{\boldsymbol{q}}} \dot{\boldsymbol{q}} - \frac{\partial \mathcal{L}(\boldsymbol{q}, \dot{\boldsymbol{q}})}{\partial \boldsymbol{q}}, \\
&= \boldsymbol{H} \ddot{\boldsymbol{q}} + \dot{\boldsymbol{H}} \dot{\boldsymbol{q}} - \frac{1}{2}\left(\dot{\boldsymbol{q}}^{\top} \frac{\partial \mathbf{H}}{\partial \boldsymbol{q}} \dot{\boldsymbol{q}}\right)^{\top} + \frac{\partial V}{\partial \boldsymbol{q}}.
\end{aligned}
\tag{3.11}
$$

The partial derivatives within the forward and inverse model can be computed using automatic differentiation or symbolic differentiation. See [68] for the symbolic differentiation of the mass matrix and the deep networks.

The system energy cannot be learned using supervised learning as the system energy cannot be observed. Therefore, the network weights of the kinetic and potential energy are learned unsupervised using the temporal consequences of the actions and system energy. One approach to learn the network parameters is to minimize the residual of the Euler–Lagrange differential equation. This optimization problem is described by

$$\boldsymbol{\psi}^*, \boldsymbol{\phi}^* = \arg\min_{\boldsymbol{\psi}, \boldsymbol{\phi}} \left\| \frac{d}{dt} \frac{\partial \mathcal{L}}{\partial \dot{\boldsymbol{q}}} - \frac{\partial \mathcal{L}}{\partial \boldsymbol{q}_i} - \boldsymbol{\tau} \right\|^2_{\boldsymbol{W}_\tau}, \tag{3.12}$$

with the Mahalanobis norm $\|\cdot\|_{\boldsymbol{W}}$ and the diagonal covariance matrix of the generalized forces $\boldsymbol{W}_\tau$. It is beneficial to normalize the loss using the covariance matrix because magnitude of the residual might vary between different joints. This optimization can be solved using any gradient-based optimization technique. Minimizing the squared residual is equivalent to fitting the inverse model, i.e., $\boldsymbol{\psi}^*, \boldsymbol{\phi}^* = \arg\min_{\boldsymbol{\psi}, \boldsymbol{\phi}} \left\| \boldsymbol{\tau} - f^{-1}(\boldsymbol{q}, \dot{\boldsymbol{q}}, \ddot{\boldsymbol{q}};\ \boldsymbol{\psi}, \boldsymbol{\phi}) \right\|^2_{\boldsymbol{W}_\tau}$. This loss can also be extended to include the forward prediction that fits the joint accelerations. The combined optimization problem is described by

$$\begin{aligned} \boldsymbol{\psi}^*, \boldsymbol{\phi}^* = \arg\min_{\boldsymbol{\psi}, \boldsymbol{\phi}} & \left\| \boldsymbol{\tau} - f^{-1}(\boldsymbol{q}, \dot{\boldsymbol{q}}, \ddot{\boldsymbol{q}};\ \boldsymbol{\psi}, \boldsymbol{\phi}) \right\|^2_{\boldsymbol{W}_\tau} \\ & + \|\ddot{\boldsymbol{q}} - f(\boldsymbol{q}, \dot{\boldsymbol{q}}, \boldsymbol{\tau};\ \boldsymbol{\psi}, \boldsymbol{\phi})\|^2_{\boldsymbol{W}_{\ddot{q}}}, \end{aligned} \tag{3.13}$$

with the diagonal covariance matrix of the generalized forces $\boldsymbol{W}_\tau$ and accelerations $\boldsymbol{W}_{\ddot{q}}$. Furthermore, its beneficial to add regularization in the form of weight decay as the Lagrangian is not unique. The Euler–Lagrange equation is invariant to linear transformation. Hence, the Lagrangian $\mathcal{L}' = \alpha L + \beta$ solves the Euler–Lagrange equation if α is non-zero and $\mathcal{L}$ is a valid Lagrangian. Therefore, adding weight regularization helps obtaining a unique solution.

3.4.1.1 Positive-Definite Mass Matrix

To obtain a physically plausible kinetic energy, the mass matrix has to be positive definite, i.e.,

$$\boldsymbol{q}^\top \boldsymbol{H}(\boldsymbol{q})\, \boldsymbol{q} > 0 \quad \forall \quad \boldsymbol{q} \in \mathbb{R}^n_0. \tag{3.14}$$

This constraint ensures all non-zero velocities have positive kinetic energy for all joint configurations. We obtain a positive definite mass matrix by predicting the Cholesky decomposition of the mass matrix with a small positive offset ϵ on the diagonal instead of the mass matrix directly. Therefore, the mass matrix is described by

$$\boldsymbol{H}(\boldsymbol{q}) = \boldsymbol{L}(\boldsymbol{q})\boldsymbol{L}(\boldsymbol{q})^{\top} + \epsilon \boldsymbol{I}, \tag{3.15}$$

with lower triangular matrix $\boldsymbol{L}$ and identity matrix $\boldsymbol{I}$. In addition, the diagonal needs to be positive as otherwise, the mass matrix is only positive semi-definite. A positive diagonal is ensured by adding a non-negative linearity to the elements of the diagonal and the positive offset ϵ. Using this parametrization the mass matrix is ensured to be positive definite for all joint configurations. However, this parametrization is numerically not favorable because the random weights the diagonal is close to ϵ for all inputs. This small diagonal is problematic for the forward model as the small eigenvalues of the mass matrix lead to a large amplification of the control torques due to the matrix inverse. The default diagonal can be shifted by adding the positive constant α before the non-linearity. This shift is described by

$$\boldsymbol{l}_{\text{diag}} = \sigma(\boldsymbol{l}_{\text{diag}} + \alpha) + \epsilon,$$

with the vectorized diagonal $\boldsymbol{l}_{\text{diag}}$ and the softplus function σ. If $\alpha > 1$, the mass matrix damps the applied torques when $\boldsymbol{l}_{\text{diag}} \approx \boldsymbol{0}$. This transformation is not essential to obtain good results but balances the forward and inverse losses.

3.4.1.2 Advantages of DeLaN

In contrast to the black-box model, this parametrization of the dynamics has three advantages, (1) this approach yields a physically plausible model that conserves energy, (2) is interpretable, and (3) can be used as forward, inverse, and energy model. The DeLaN model is guaranteed to evolve like a mechanical system and is passive [92] as the forward dynamics are derived from the physics prior and the positive definite mass matrix for all model parameters. If the system is uncontrolled, i.e., $\boldsymbol{\tau} = \boldsymbol{0}$, the system energy is conserved as the change in energy is described by $\dot{\mathcal{H}} = \dot{\boldsymbol{q}}^{\top}\boldsymbol{\tau} = \boldsymbol{0}$. In contrast, black-box models can generate additional energy without system inputs. It is important to note that the conservation of energy of DeLaN does not guarantee to prevent the divergence of the model rollouts. The potential energy is not bounded and can accelerate the system indefinitely. Especially outside the training domain, the potential energy is random.

The model is interpretable as one can disambiguate between the different forces, e.g., inertial-, centrifugal-, Coriolis, and gravitational force. This decomposition is beneficial as some model-based control approaches require the explicit computation of the mass matrix, the gravitational force, or the system energy. Furthermore, the same model parameters can be used for the forward, inverse, and energy model. Therefore, the forward and inverse model are consistent. In contrast, black-box models need to learn separate parameters for the inverse and forward model that might not be consistent and cannot obtain the system energy as these cannot be observed.

3.4.2 Hamiltonian Neural Networks (HNN)

Instead of using Lagrangian Mechanics as model prior for deep networks, [33] proposed to use Hamiltonian mechanics. In this case, the HNN parametrize the Hamiltonian with two deep networks described by

$$\mathcal{H}(\boldsymbol{q}, \boldsymbol{p};\ \boldsymbol{\psi}, \boldsymbol{\phi}) = \frac{1}{2}\boldsymbol{p}^{\top}\boldsymbol{H}(\boldsymbol{q};\ \boldsymbol{\psi})^{-1}\,\boldsymbol{p} - V(\boldsymbol{q};\ \boldsymbol{\phi}). \tag{3.16}$$

It is important to note that HNN predict the inverse of the mass matrix instead of the mass matrix as in DeLaN. Similar to DeLaN, the forward and inverse model can be derived. The forward model $\left[\dot{\boldsymbol{q}}, \dot{\boldsymbol{p}}\right] = f(\boldsymbol{q}, \boldsymbol{p}, \boldsymbol{\tau};\ \boldsymbol{\psi}, \boldsymbol{\phi})$ is described by

$$\dot{\boldsymbol{q}} = \frac{\partial \mathcal{H}(\boldsymbol{q}, \boldsymbol{p})}{\partial \boldsymbol{p}} = \boldsymbol{H}^{-1}\boldsymbol{p},$$

$$\dot{\boldsymbol{p}} = -\frac{\partial \mathcal{H}(\boldsymbol{q}, \boldsymbol{p})}{\partial \boldsymbol{q}} + \boldsymbol{\tau} = -\frac{1}{2}\left(\boldsymbol{p}^{\top}\frac{\partial \boldsymbol{H}^{-1}}{\partial \boldsymbol{q}}\boldsymbol{p}\right)^{\top} - \frac{\partial V}{\partial \boldsymbol{q}} + \boldsymbol{\tau}.$$

The inverse model $\boldsymbol{\tau} = f^{-1}(\boldsymbol{q}, \boldsymbol{p}, \dot{\boldsymbol{p}};\ \boldsymbol{\psi}, \boldsymbol{\phi})$ is described by

$$\begin{aligned}\boldsymbol{\tau} &= \dot{\boldsymbol{p}} + \frac{\partial \mathcal{H}(\boldsymbol{q}, \boldsymbol{p})}{\partial \boldsymbol{q}},\\ &= \dot{\boldsymbol{p}} - \frac{1}{2}\left(\boldsymbol{p}^{\top}\boldsymbol{H}^{-1}\frac{\partial \boldsymbol{H}}{\partial \boldsymbol{q}}\boldsymbol{H}^{-1}\boldsymbol{p}\right)^{\top} + \frac{\partial V}{\partial \boldsymbol{q}}.\end{aligned}$$

The network parameters of the kinetic and potential energy can be obtained by minimizing the squared residual using the observed data consisting of $\left[\boldsymbol{q}, \boldsymbol{p}, \dot{\boldsymbol{q}}, \dot{\boldsymbol{p}}, \boldsymbol{\tau}\right]$. This optimization is described by

$$\begin{aligned}\boldsymbol{\psi}^{*}, \boldsymbol{\phi}^{*} = \arg\min_{\psi,\phi} &\left\| \dot{\boldsymbol{p}} + \frac{\partial \mathcal{H}(\boldsymbol{q}, \boldsymbol{p})}{\partial \boldsymbol{q}} - \boldsymbol{\tau} \right\|^{2}_{\boldsymbol{W}_{\dot{p}}}\\ &+ \left\| \dot{\boldsymbol{q}} - \frac{\partial \mathcal{H}(\boldsymbol{q}, \boldsymbol{p})}{\partial \boldsymbol{p}} \right\|^{2}_{\boldsymbol{W}_{\dot{q}}},\end{aligned} \tag{3.17}$$

with the diagonal covariance matrix $\boldsymbol{W}_{\dot{q}}$ and $\boldsymbol{W}_{\dot{p}}$ of $\dot{\boldsymbol{q}}$ and $\dot{\boldsymbol{p}}$. The minimization can be solved using the standard gradient based optimization toolkit and automatic differentiation.

3.4.2.1 Differences to DeLaN

DeLaN and HNN share the same advantages as both models are derived from the same principle. Therefore, HNN conserve energy, are interpretable, and provide a forward, inverse, and energy model. The main difference is that DeLaN uses position and velocity while HNN uses position and momentum. Depending on the observed quantities either model fits better than the other. A minor difference is that minimizing the residual of the Euler–Lagrange equation is identical to the inverse model loss while minimizing the residual of Hamilton's equations is identical to the forward model loss.

From a numerical perspective, the Hamiltonian mechanics prior is slightly beneficial as the forward and inverse model only relies on the inverse of the mass matrix. Therefore, one does not need to numerically compute the inverse of the predicted matrix. Avoiding the explicit inversion makes the learning and model rollout a bit more stable. The Lagrangian mechanics prior relies on the mass matrix as well on the inverse. Therefore, the inverse of the predicted matrix has to be computed numerically. When the eigenvalues of the mass matrix approach ϵ and $\epsilon \ll 1$, the model rollout and the optimization of the forward model can become numerically sensitive. Therefore, it is important to choose ϵ as large as possible for the corresponding system as this limits the amplification of the acceleration.

3.4.3 *Variations of DeLaN and HNN*

Since the introduction of DeLaN [68] and HNN [33], many other variants and extensions have been proposed within the literature. We provide an overview of the existing work and highlight the differences.

3.4.3.1 Parametrization of $\mathcal{L}$ and $\mathcal{H}$

In the previous sections, the Hamiltonian $\mathcal{H}$ and Lagrangian $\mathcal{L}$ were parameterized by two networks predicting the mass matrix, or its inverse, for the kinetic energy and the potential energy. Instead of predicting these two quantities separately, one can also use a single feed-forward network for both quantities. This factorization is described by

$$\mathcal{L} = h(\boldsymbol{q}, \dot{\boldsymbol{q}};\ \boldsymbol{\psi}), \qquad\qquad \mathcal{H} = h(\boldsymbol{q}, \boldsymbol{p};\ \boldsymbol{\psi}),$$

with the standard feed-forward network h and the network parameters $\boldsymbol{\psi}$. Within the literature, this approach was used by [21, 33] while the [26, 34, 68, 86, 106] used the representation of kinetic and potential energy. In the following we, will differentiate between both approaches by using the term structured Lagrangian/Hamiltonian and black-box Lagrangian/Hamiltonian. The structured approach, represent the mass matrix and potential energy explicitly while the black-box approach uses a single network to represent the Lagrangian or Hamiltonian. The differences in the model architecture for both approaches are depicted in Fig. 3.2.

One benefit of using a black-box $\mathcal{L}$ and $\mathcal{H}$ is that the quadratic parametrization of the kinetic energy does not apply to relativistic systems. The disadvantages of a single network approach are that this parametrization is computationally more demanding as one needs to compute the Hessian of the network. Evaluating the Hessian of a deep network can be done using automatic differentiation, but is expensive in terms of computation and memory. When using the quadratic kinetic energy, computing the Hessian of the network is not needed. Furthermore, the Hessian may not be invertible if only a single network is used. If the Hessian is singular or nearly singular, the forward model using the Lagrangian prior becomes unstable and diverges. For structured Lagrangian, this problem does not occur as the eigenvalues of the mass matrix are lower bounded.

Most existing work uses standard feed-forward networks to model the system energy, the Hamiltonian or the Lagrangian [26, 33, 34, 68, 86, 106]. Other variants have also applied the physics-inspired networks to graph neural networks [10, 21, 87]. Such graph neural networks incorporate additional structure within the network architecture when the system dynamics consist of multiple identical particles without additional constraints. Therefore, these methods exhibit improved performance for modeling N-body problems.

3.4.3.2 Loss Functions and Integrators

The loss functions of DeLaN (Eq. 3.13) and HNN (Eq. 3.17) express the loss in terms including the acceleration, i.e., $\ddot{\boldsymbol{q}}$ and $\dot{\boldsymbol{p}}$. These quantities are commonly not observed for real-world systems and are approximated using finite differences. The problem of this approximation is that the finite differences amplify the amplitude of high frequency noise components. Therefore, one has to use low-pass filters to obtain good acceleration estimates. A different approach that avoids approximating the accelerations is to only use the forward loss and reformulate the loss in terms of position and velocities. In this case, the loss is described by

$$\boldsymbol{\psi}^*, \boldsymbol{\phi}^* = \arg\min_{\boldsymbol{\psi}, \boldsymbol{\phi}} \left\| \boldsymbol{x}_{t+1} - \hat{\boldsymbol{x}}_{t+1}(\boldsymbol{x}_t, \boldsymbol{\tau}_t; \boldsymbol{\psi}, \boldsymbol{\phi}) \right\|^2, \tag{3.18}$$

with the predicted next state $\hat{\boldsymbol{x}}$, the state $\boldsymbol{x} = [\boldsymbol{q}, \dot{\boldsymbol{q}}]$ in the case of Lagrangian formulation and the state $\boldsymbol{x} = [\boldsymbol{q}, \boldsymbol{p}]$ in the Hamiltonian formulation. The predicted next state can be obtained by solving the differential equation

$$\begin{bmatrix} \dot{\boldsymbol{q}} \\ \ddot{\boldsymbol{q}} \end{bmatrix} = \begin{bmatrix} \dot{\boldsymbol{q}} \\ \frac{\partial^2 \mathcal{L}(\boldsymbol{q}, \dot{\boldsymbol{q}})}{\partial^2 \dot{\boldsymbol{q}}}^{-1} \left[\boldsymbol{\tau} - \frac{\partial \mathcal{L}(\boldsymbol{q}, \dot{\boldsymbol{q}})}{\partial \boldsymbol{q} \partial \dot{\boldsymbol{q}}} \dot{\boldsymbol{q}} + \frac{\partial \mathcal{L}(\boldsymbol{q}, \dot{\boldsymbol{q}})}{\partial \boldsymbol{q}} \right] \end{bmatrix},$$

$$\begin{bmatrix} \dot{\boldsymbol{q}} \\ \dot{\boldsymbol{p}} \end{bmatrix} = \begin{bmatrix} \frac{\partial \mathcal{H}(\boldsymbol{q}, \boldsymbol{p})}{\partial \boldsymbol{p}} \\ -\frac{\partial \mathcal{H}(\boldsymbol{q}, \boldsymbol{p})}{\partial \boldsymbol{q}} + \boldsymbol{G}(\boldsymbol{q})\, \boldsymbol{\tau} \end{bmatrix},$$

using any numerical integration approach. In the case of the explicit Euler integration this approach is identical to the loss of Eqs. 3.13 and 3.17. A common choice to compute the next step is the Runge Kutta 4 (RK4) fixed time step integrator [33, 35]. This loss formulation also enables a multi-step loss which has been shown to improve the performance of model predictive control for deterministic models [66]

A more elaborate approach has been proposed by [86] that combines discrete mechanics with variational integrators. This combination guarantees that even the discrete-time system conserves momentum and energy. The RK4 integration might leak or add energy due to the discrete-time approximation. The main disadvantage of the variational integrator networks is that this approach assumes a constant mass matrix. Therefore, the Coriolis and centrifugal force disappear (Eq. 3.6) and the acceleration only depends on the position. Within the discrete mechanics literature, extensions exist to apply the variational integrator to multi-body systems with a non-constant mass matrix. However, these extensions are non-trivial and involve solving a root-finding problem within each integration step [61].

3.4.3.3 Feature Transformation

The previous sections always used generalized coordinates or momentum to describe the system dynamics. However, this formulation can be problematic as these coordinates are unknown or unsuitable for function approximation. For example, continuous revolute joints without angular limits are problematic for function approximation due to the wrapping of the angle at $\pm\pi$. This problem is commonly mitigated using sine/cosine feature transformations. Such feature transformation can be included in physics-inspired networks if the feature transforms mapping from the generalized coordinates to the features z is known and differentiable. The more general problem of only observing the features and unknown feature transformation and generalized coordinates is discussed in Sect. 3.6.1.2.

Let g be the feature transform mapping generalized coordinates to the features $z = g(\boldsymbol{q})$. For continuous revolute joints the feature transform $\boldsymbol{g}(\boldsymbol{q}) = \left[\cos \boldsymbol{q}_0, \sin \boldsymbol{q}_0\right]$ avoids the problems associated with wrapping the angle. In this case the Lagrangian is described by

$$\mathcal{L}(z, \dot{\boldsymbol{q}};\ \boldsymbol{\psi}, \boldsymbol{\phi}) = \frac{1}{2}\dot{\boldsymbol{q}}^\top \boldsymbol{H}(z;\ \boldsymbol{\psi})\, \dot{\boldsymbol{q}} - V(z;\ \boldsymbol{\phi}).$$

In this case, one can apply the chain rule to obtain the gradients w.r.t. the generalized coordinates, i.e.,

$$\frac{\partial \mathcal{L}}{\partial \boldsymbol{q}} = \frac{\partial \boldsymbol{g}}{\partial \boldsymbol{q}}^\top \frac{\partial \mathcal{L}}{\partial z}.$$

This approach is identical to adding an input layer to the neural network with the hand-crafted transformations. The feature transformation was previously introduced by [106]. However, the authors only manually derived the special case for continuous angle while this approach can be easily generalized to arbitrary differentiable feature transformations.

3.4.3.4 Actuator Models and Friction

The physics-inspired networks cannot model friction directly as the learned dynamics are conservative. Incorporating friction within this model learning approach in a non-black-box fashion is non-trivial because friction is an abstraction to combine various physical effects. For robot arms in free space, the friction of the motors dominates, for mechanical systems dragging along a surface the friction at the surface dominates while for legged locomotion the friction between the feet and floor dominates but also varies with time. Therefore, defining a general case for all types of friction in compliance with the Lagrangian and Hamiltonian Mechanics is challenging. Various approaches to incorporate friction models analytically can be found in [65, 101].

Most existing works on physics-inspired networks only focus on friction caused by the actuators, which dominates for robot arms [35, 67, 69]. In this case the friction can be expressed using generalized coordinates and is a non-conservative force. Incorporating other types of friction than actuator friction is non-trivial as these cannot be easily expressed using the generalized force. In this case, one requires the contact-point and contact Jacobian to map the contact-force to the generalized force. For the actuator model, the generalized force required for DeLaN and HNN is expressed using an addition function that modulates the system input and adds friction. This function is described by $\boldsymbol{\tau} = g(\boldsymbol{q}, \dot{\boldsymbol{q}}, \boldsymbol{u})$ with the system input $\boldsymbol{u}$. For the actuator model, one can either choose a white-box approach that uses a analytic actuator and friction models [67] or a black-box approach that uses a deep network [35, 106]. For example, a white-box model can add a friction torque $\boldsymbol{\tau}_f$ to motor torque $\boldsymbol{u}$. Therefore, the generalized force is described $\boldsymbol{\tau} = \boldsymbol{u} + \boldsymbol{\tau}_f$. Within the literature many friction models have been proposed [2, 8, 77, 97]. These models assume that the motor friction only depends on the joint velocity $\dot{\boldsymbol{q}}_i$ of the ith-joint and is independent of the other joints. Common choices for friction are static, viscous, stiction described by

$$\begin{aligned} &\text{Coulomb Friction} && \boldsymbol{\tau}_f = -\boldsymbol{\tau}_c, \\ &\text{Viscous Friction} && \boldsymbol{\tau}_f = -\boldsymbol{\rho} \odot \dot{\boldsymbol{q}}, \\ &\text{Stiction} && \boldsymbol{\tau}_f = -\tau_s \odot sign\,(\dot{\boldsymbol{q}}) \odot \exp\left(-\,\dot{\boldsymbol{q}}^2/\nu\right), \end{aligned}$$

with the elementwise multiplication $\odot$, the Coulomb friction constant $\boldsymbol{\tau}_c$, the viscous friction constant $\boldsymbol{\rho}$ and the stiction constants $\boldsymbol{\tau}_s$ and ν. These friction models can also be combined to yield the Stribeck friction described by

$$\boldsymbol{\tau}_f = -\left(\tau_c + \tau_s \odot \exp\left(-\dot{\boldsymbol{q}}^2/\nu\right)\right) \odot \operatorname{sign}(\dot{\boldsymbol{q}}) - \boldsymbol{d} \odot \dot{\boldsymbol{q}}.$$

It is important to note that the system is not time-reversible when stiction is added to the dynamics as multiple motor-torques can generate the same joint acceleration [82].

In contrast to these white-box approaches, [35, 106] proposed to add an black-box actuator model. For example, [35] proposed to use a black-box control matrix $\boldsymbol{G}(\boldsymbol{q})$ with viscous friction for DeLaN. Therefore, the actuator model is described by

$$\boldsymbol{\tau} = \boldsymbol{G}(\boldsymbol{q}, \dot{\boldsymbol{q}}; \boldsymbol{\theta})\, \boldsymbol{u} - \boldsymbol{\rho} \odot \dot{\boldsymbol{q}}, \tag{3.19}$$

with the positive friction coefficients $\boldsymbol{\rho}$. The control matrix $\boldsymbol{G}$ is predicted by an additional neural network. Similarly, [107] proposed to use a state-dependent control matrix $\boldsymbol{G}(\boldsymbol{q})$ and a positive definite dissipation matrix $\boldsymbol{D}(\boldsymbol{q})$ for HNN. In this case the generalized force is described by

$$\boldsymbol{\tau} = \boldsymbol{G}(\boldsymbol{q})\, \boldsymbol{u} - \boldsymbol{D}(\boldsymbol{q}) \begin{bmatrix} \frac{\partial \mathcal{H}(\boldsymbol{q}, \boldsymbol{p})}{\partial \boldsymbol{q}} \\ \frac{\partial \mathcal{H}(\boldsymbol{q}, \boldsymbol{p})}{\partial \boldsymbol{p}} \end{bmatrix}.$$

Both matrices are predicted using a deep network. The network parameters of the actuator model are optimized using gradient descent. These black-box actuator models can represent more complex actuator dynamics and even system dynamics violating the assumptions of Lagrangian and Hamiltonian Mechanics. However, this actuator model can also result that the potential and kinetic energy are ignored and only the black-box model dominates the predicted dynamics. To avoid that the actuator model predicts the complete system dynamics, it is beneficial to add penalties to the magnitude of the actuator during the optimization. The existing grey-box model learning literature [4, 49, 69] has shown that these penalties improve the performance.

3.5 Experiments

In the experiments, we apply physics-inspired deep network models to learn the non-linear dynamics of simulated systems and physical systems. Within the simulation experiments, we want to test whether the different physics-inspired networks learn the underlying structure and highlight the empirical differences of the existing approaches. On the physical systems, we compare the model-based control performance of DeLaN with a structured Lagrangian for the fully-actuated and under-actuated system to standard system identification techniques and black-box model learning. We only use DeLaN for the physical systems as for these systems we do not observe the momentum. Hence, only the Lagrangian physics prior is applicable. One could treat the Hamiltonian prior as a latent space problem with the momentum being the latent representation. However, this approach would effectively boil down to the Lagrangian prior. Using these experiments, we want to answer the following questions:

Q1: Do physics-inspired networks learn the underlying representation of the dynamical system?

Q2: Do physics-inspired networks perform better than continuous-time black-box models?

Q3: Can physics-inspired networks be applied to physical systems where the physics prior does not hold?

3.5.1 Experimental Setup

To answer these questions, we apply the different variations of physics-inspired models to 4 different systems and compare the performance to three baselines. Within the experiments, we denote the physics-inspired networks that only use a single network to represent the Lagrangian or Hamiltonian as black-box DeLaN/HNN. When two separate networks are used to represent that mass-matrix and potential energy, we refer to this approach as structured DeLaN/HNN. The detailed differences between both approaches are described in Sect. 3.4.3.1. For each of the experiments the dynamics models are learned from a fixed dataset and are trained until convergence. All evaluations are performed on a test dataset that is not contained within the training dataset.

In the following, we briefly introduce the systems and baselines. The code of Deep Lagrangian Networks (DeLaN) and Hamiltonian Neural Networks (HNN) is available at https://github.com/milutter/deep_lagrangian_networks.

3.5.1.1 Plants

Within the experiments, we apply the model learning techniques to a simulated two-link pendulum, the Barrett WAM, the cartpole, and the Furuta pendulum. For all physical systems, only the joint position is directly observed. The velocities and accelerations are computed using finite differences and low-pass filters. These filters are applied offline to use zero-phase shift filters that do not cause a phase shift within the observations. The hyperparameters of DeLaN for each system are described in Table 3.1.

Two-link Pendulum. The two-link pendulum has two continuous revolute joints, is fully actuated, and acts in the vertical x-z plane with gravity. The pendulum is simulated using Bullet [20].

Cartpole. The physical cartpole (Fig. 3.3a) is an under-actuated system manufactured by [79]. The pendulum is passive and the cart is voltage controlled

Table 3.1 The hyperparameters of DeLaN used for the different dynamical systems. All physical system have a fixed sampling time of 2.0 ms. The number of training samples for the 2-DoF robot differs by dataset. See Table 3.2 for more details about the different datasets

	2-DoF robot	Cartpole	Furuta pendulum	Barrett WAM
Total DoF/actuated DoF	2/2	2/1	2/1	4/4
Number of training samples	2500 & 10^5	10^5/≈200 s	10^5/≈200 s	10^5/≈200 s
Network dimension	[2 × 64]	[2 × 256]	[2 × 128]	[2 × 128]
Activation	Tanh	SoftPlus	SoftPlus	SoftPlus
Batch size	512	1024	1024	1024
Learning rate	10^{-4}	10^{-4}	10^{-4}	10^{-4}
Weight decay	10^{-5}	10^{-5}	10^{-5}	10^{-5}
Optimizer	ADAM	ADAM	ADAM	ADAM

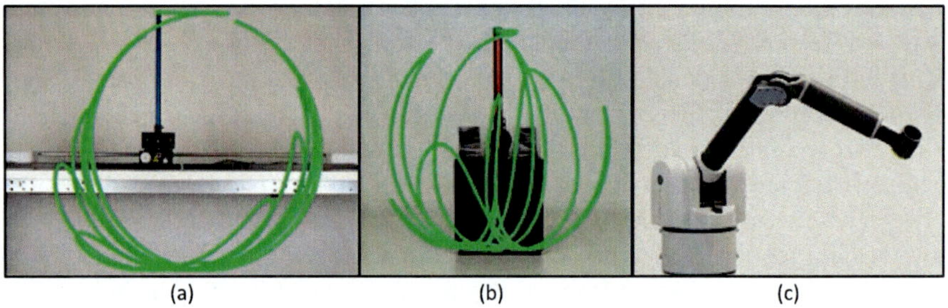

Fig. 3.3 The **a** Cartpole, **b** Furuta pendulum and **c** Barrett WAM used for the evaluation. The Furuta pendulum and cartpole perform a swing-up using the energy controller. The Barrett WAM executes a cosine trajectory with a different frequency per joint

with up to 500 Hz. The linear actuator consists of a plastic cogwheel drive with high stiction.

Furuta Pendulum. The physical Furuta pendulum (Fig. 3.3b) is an under-actuated system manufactured by [79]. Instead of the linear actuator of the cartpole, the Furuta pendulum has an actuated revolute joint and a passive pendulum. The revolute joint is voltage controlled with up to 500 Hz. The main challenge with this system is the small masses and length scale of the system. These characteristics yield a very sensitive control system.

Barrett WAM. The Barrett WAM (Fig. 3.3c) consists of four actuated degrees of freedom controlled via torque control with 500 Hz. The actuators are back-driveable and consist of cable drives with low gear ratios enabling fast accelerations. The joint angles sensing is on the motor-side. Therefore, any deviation between the motor position and joint position due to the slack of the cables cannot be observed. We only use the 4 degree of freedom version as the wrist and end-effector joints cannot be excited due to the limited range of motion and acceleration.

3.5.1.2 Baselines

We use the analytic dynamics model, system identification, and a feed-forward deep network as baselines.

Analytic Model. The analytic model uses the equation of motion derived using rigid body dynamics and the system parameters, i.e., masses, center of gravity, and inertias, provided by the manufacturer. In addition to the rigid body dynamics, these models are augmented with a viscous friction model.

System Identification. This approach requires the knowledge of the analytic equations of motions and infers the system parameters from data. More specifically we use the technique described by [6]. This approach showed that for rigid body kinematic trees the inverse dynamics model is a linear model described by

$$\boldsymbol{\tau} = \boldsymbol{A}(\boldsymbol{q}, \dot{\boldsymbol{q}}, \ddot{\boldsymbol{q}})\, \boldsymbol{\theta}, \tag{3.20}$$

with the with hand-crafted features $\boldsymbol{A}(\cdot)$ derived from the kinematics and the system parameters $\boldsymbol{\theta}$. As the inverse dynamics are a linear model, the system parameters can be obtained using linear regression. We additionally penalize deviations from the parameters nominal parameters provided by the manufacturer. In this case the optimal parameters inferred from data are obtained by

$$\boldsymbol{\theta}^* = \boldsymbol{\theta}_0 + \left(\boldsymbol{A}^\top \boldsymbol{A} + \lambda^2 \boldsymbol{I}\right)^{-1} \boldsymbol{A}^\top \Big(\boldsymbol{\tau} - \boldsymbol{A}\boldsymbol{\theta}_0\Big), \tag{3.21}$$

with the nominal parameters $\boldsymbol{\theta}_0$ and the regularization constant λ. The resulting system parameters might not be physically plausible as the individual elements of $\boldsymbol{\theta}$ have additional constraints [94]. For example the masses have to be positive and the inertias have to adhere to the parallel axis theorem.

Feed-Forward Network. The deep network baseline uses two separate networks, where one describes the forward dynamics and the other the inverse dynamics. This model does not necessarily generate coherent predictions as the parameters of the forward and inverse model are decoupled. Therefore, it is not guaranteed that $f(f^{-1}(\boldsymbol{x})) = \boldsymbol{x}$ holds. The forward model is a continuous-time model and predicts the joint acceleration $\ddot{\boldsymbol{q}}$. Therefore, the deep network baseline is independent of the sampling frequency and uses an explicit integrator as the physics-inspired network. The network parameters are learned by minimizing the normalized squared error of the forward and inverse model. This optimization problem is solved by gradient descent using ADAM. This baseline cannot be applied to the energy experiments, as the system energy cannot be learned by a standard deep network.

3.5.2 Model Prediction Experiments

For the simulated experiments, we want to evaluate whether the physics-inspired networks can learn the underlying system dynamics and recover the structure with ideal observations. Therefore, we want to observe the data fit and as well as the long-term forward predictions. Furthermore, we want to differentiate between two separate datasets, (1) a large data set with 100 k samples spanning the state domain uniformly and (2) a small dataset with only 2.5 k samples which consist of trajectories of drawing characters. For the character test dataset, the test set contains different characters than the training set. This dataset only spans a small sub-domain of the state space. The character dataset was initially introduced by [104] and is available in the UCI Machine Learning Repository [23]. For training, the datasets are split into a test and training set. The reported results are reported on the test set and averaged over 5 seeds.

3.5.2.1 Inverse Model

The results of the inverse model are summarized in Table 3.2 and visualized in Fig. 3.4. All models learn a good inverse model that fits the test set. When comparing the performance across the large and small datasets one cannot observe a difference in performance. All models perform comparably for the small and large datasets. On average, the physics-inspired networks obtain a lower MSE than the black-box deep network. When comparing the structured Lagrangian/Hamiltonian to the black-box counterparts no clear difference is observable for the inverse model.

When comparing the torque decomposition of the inertial, centrifugal, Coriolis, and gravitational forces, all models learn a good decomposition. For the unstructured models, this decomposition can be evaluated by assuming the underlying structure and evaluating the inverse model. For example, the gravitational component by evaluating $\boldsymbol{\tau}_g = f^{-1}(\boldsymbol{q}, \mathbf{0}, \mathbf{0})$. All models learn the underlying structure that fits the true decomposition. Even the black-box feed-forward network obtains a good decomposition despite having no structure. When comparing the MSE error in Table 3.2, the MSE for the physics-inspired networks is better than the black-box feed-forward network. This difference is especially pronounced for the inertial and Coriolis torque. The difference in the gravitational torque is not so large. When comparing the decomposition of the black-box Lagrangian/Hamiltonian to the structured counterparts, the structured approach outperforms the black-box approach on the inertial and Coriolis torque.

3.5.2.2 Forward Model

The results of the forward model are summarized in Table 3.2 and visualized in Fig. 3.5. Also for the forward model, the physics-inspired networks obtain a better

Table 3.2 The normalized mean squared error (nmse) and mean valid prediction time (VPT) as well as the corresponding confidence interval averaged over 10 seeds. On average the structured Hamiltonian and Lagrangian approaches obtain better forward and inverse models than the black-box counterparts and the standard feed forward neural network. When observing the corresponding phase space coordinates, the Hamiltonian and Lagrangian approaches perform comparable

Uniform data		Inverse model			
		Torque—$\boldsymbol{\tau}$	Inertial torque $\boldsymbol{\tau}_I$	Coriolis torque $\boldsymbol{\tau}_c$	Gravitational torque $\boldsymbol{\tau}_g$
DeLaN	Structured	**2.2e−7 ± 3.2e−6**	**2.9e−9 ± 4.5e−8**	**2.5e−8 ± 3.0e−7**	**1.0e−8 ± 3.2e−8**
DeLaN	Black-Box	2.1e−4 ± 4.9e−3	**4.0e−9 ± 2.1e−8**	1.9e−5 ± 3.9e−4	**3.0e−8 ± 7.2e−8**
HNN	Structured	**4.6e−7 ± 1.9e−6**	**4.6e−9 ± 2.8e−8**	**8.1e−8 ± 5.3e−7**	**3.5e−8 ± 8.3e−8**
HNN	Black-Box	3.3e−5 ± 5.5e−4	9.9e−6 ± 6.0e−5	3.3e−5 ± 3.1e−4	**5.9e−8 ± 1.4e−7**
FF-NN	–	5.8e−5 ± 1.0e−3	2.3e−7 ± 1.5e−6	9.9e−6 ± 1.4e−4	2.9e−7 ± 7.4e−7
Character data					
DeLaN	Structured	**7.7e−8 ± 2.1e−7**	2.0e−7 ± 1.7e−6	**1.1e−6 ± 5.2e−6**	2.0e−7 ± 1.0e−6
DeLaN	Black-Box	**9.9e−8 ± 4.4e−7**	**9.4e−8 ± 4.7e−7**	**1.7e−6 ± 1.2e−5**	1.3e−7 ± 1.1e−6
HNN	Structured	**1.8e−8 ± 5.5e−8**	**1.5e−8 ± 9.1e−8**	**5.5e−7 ± 1.9e−6**	**2.5e−8 ± 7.0e−8**
HNN	Black-Box	**5.5e−8 ± 1.8e−7**	5.0e−5 ± 2.7e−4	7.2e−4 ± 4.3e−3	**8.4e−8 ± 2.7e−7**
FF-NN	–	2.2e−6 ± 9.1e−6	4.2e−6 ± 2.5e−5	5.5e−5 ± 3.2e−4	9.2e−7 ± 5.1e−6
Uniform data		Forward model			
		State error $\dot{x}$	VPT [s]		
DeLaN	Structured	**3.9e−5 ± 5.9e−4**	**5.64s ± 1.78s**		
DeLaN	Black-Box	2.1e+1 ± 1.9e+3	3.59s ± 2.18s		
HNN	Structured	**1.1e−4 ± 4.4e−4**	**5.09s ± 1.91s**		
HNN	Black-Box	2.0e−2 ± 3.1e−1	3.80s ± 1.72s		
FF-NN	–	6.1e−3 ± 1.1e−1	2.52s ± 0.56s		
Character data					
DeLaN	Structured	**3.5e−5 ± 1.1e−4**	**2.32s ± 0.37s**		
DeLaN	Black-Box	**4.3e−5 ± 1.7e−4**	**2.76s ± 0.68s**		
HNN	Structured	**1.5e−5 ± 3.6e−5**	**2.90s ± 0.68s**		
HNN	Black-Box	6.1e−5 ± 1.6e−4	2.19s ± 0.62s		
FF-NN	–	6.5e−4 ± 3.0e−3	1.81s ± 0.49s		

performance on the state error than the feed-forward network. All models perform better on the small character dataset than on the large dataset as the state domain is much smaller than the uniform domain of the large dataset. For the large dataset, the black-box Lagrangian/Hamiltonian approaches are much worse compared to the structured counterparts. This is especially visible for the black-box Lagrangian. The average error and variance is so large because the mass matrix becomes nearly singular for some samples. The nearly singular mass matrix amplifies small differences yielding a very large error.

To compare the long-term predictions of the models, we compare the valid prediction time (VPT) [10], which is defined as the duration until the predicted rollout has

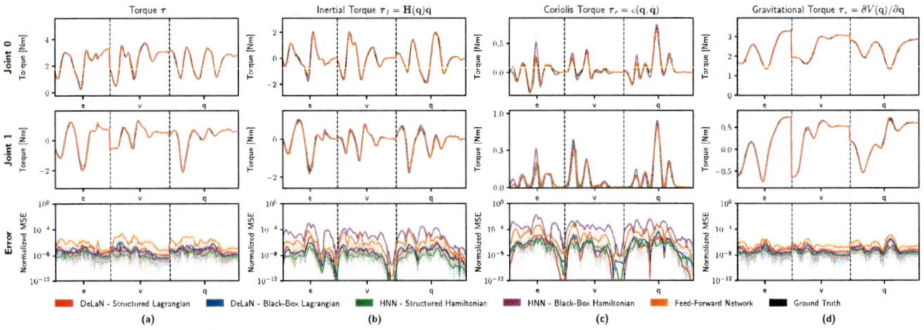

Fig. 3.4 **a** The learned inverse model using the character dataset averaged over 10 seeds. The test character 'e', 'v', 'q' are not contained within the training set. The remaining columns show predicted force decomposition. **b** Plots the inertial force $\mathbf{H}\ddot{\boldsymbol{q}}$, **c** the Coriolis and Centrifugal forces $c(\boldsymbol{q}, \dot{\boldsymbol{q}})$ and **d** the gravitational force $g(\boldsymbol{q})$. All physics-inspired networks learn a good inverse model that obtains a lower MSE than the feed-forward network. The Lagrangian approaches learn a better force decomposition than the Hamiltonian approach. This improved performance is especially visible for the inertial, centrifugal, and Coriolis torque

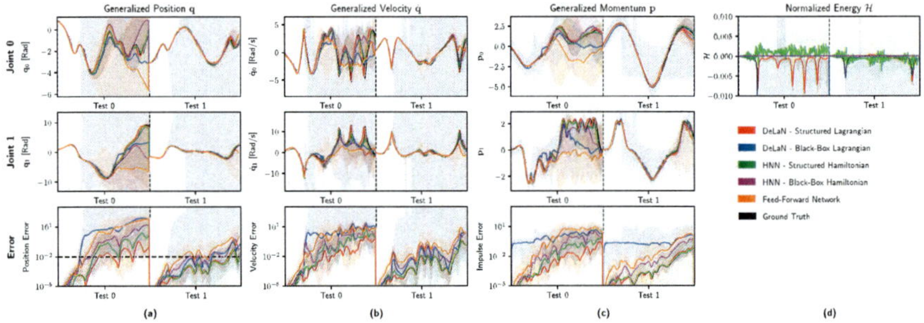

Fig. 3.5 The model rollouts of **a** the position, **b** velocity and **c** momentum, and **d** energy of the forward models for two test trajectories of the uniform dataset averaged over 10 seeds. The structured physics-inspired networks perform the best compared to the standard feed-forward network and the black-box counterparts. Especially the rollout of the black-box Lagrangian commonly diverges as the Hessian of the Lagrangian, which is required for computing the acceleration, becomes close to singular. This nearly singular Hessian causes exploding velocities and consequently also divergence of the estimated momentum computed via $\mathbf{H}\dot{\boldsymbol{q}}$

a larger error than a pre-defined threshold. We define the threshold of the MSE to be 1e−2, which corresponds to an angular error of ≈5 degrees. The long-term prediction of the physics-inspired networks is better than the prediction time of the feed-forward network on both datasets (Table 3.2). Furthermore, the structured variants of DeLaN and HNN perform better than the black-box approaches. The problem of the nearly singular mass matrix can be observed in Fig. 3.5 for the black-box Lagrangian. For one test trajectory and some seeds, the near singular mass matrix lets the trajectory diverge. For the structured HNN and DeLaN this divergence is not observed. Furthermore, the momentum prediction of the black-box DeLaN variant shows the worse accuracy of the network Hessian corresponding to the mass matrix. The predicted momentum of this approach has a much higher variance.

3.5.2.3 Conclusion

The simulated experiments show that the physics-inspired networks learn the underlying structure of the dynamical system. These models can accurately predict the force decomposition, momentum, and system energy. Furthermore, the physics-inspired models can learn better forward and inverse models than a standard feed-forward deep network. The structured DeLaN and HNN perform better than the black-box counterparts. The forward and inverse model of the structured DeLaN and HNN do not show any empirical differences when the corresponding phase space coordinates are observed.

3.5.3 Model-Based Control Experiments

With the experiments on the physical system, we want to evaluate the control performance of the learned models with noisy real-world data. Evaluating the control performance rather than the MSE on static datasets is the more relevant performance measure as the application of the models is control. Furthermore, it has been shown that the MSE is not a good substitute to predict the control performance of a learned model and commonly overestimates the performance [45, 58, 66]. To evaluate the model performance for control, we apply the learned models to inverse dynamics control and energy control. We only apply DeLaN with the structured Lagrangian to the physical systems as the potentially singular mass matrix risks damaging the physical system. HNN do not apply to the system as the momentum cannot be retrieved from the position observations while the velocity can be obtained using finite differences.

3.5.3.1 Inverse Dynamics Control

Evaluating learned models by comparing the tracking error of a model-based control law has been a well-established benchmark for evaluating the control performance of models [75, 76]. In this experiment, we use inverse dynamics control as a model-based control law. This feedback controller augments the PD-control law with an additional feed-forward torque to compensate for the non-linear dynamics of the system. Therefore, the inverse dynamics control obtains a better tracking error than the standard PD control. The resulting control law is described by

$$\boldsymbol{\tau} = \boldsymbol{K}_P \left(\boldsymbol{q}_{\text{des}} - \boldsymbol{q}\right) + \boldsymbol{K}_D \left(\dot{\boldsymbol{q}}_{\text{des}} - \dot{\boldsymbol{q}}\right) + f^{-1}(\boldsymbol{q}_{\text{des}}, \dot{\boldsymbol{q}}_{\text{des}}, \ddot{\boldsymbol{q}}_{\text{des}}),$$

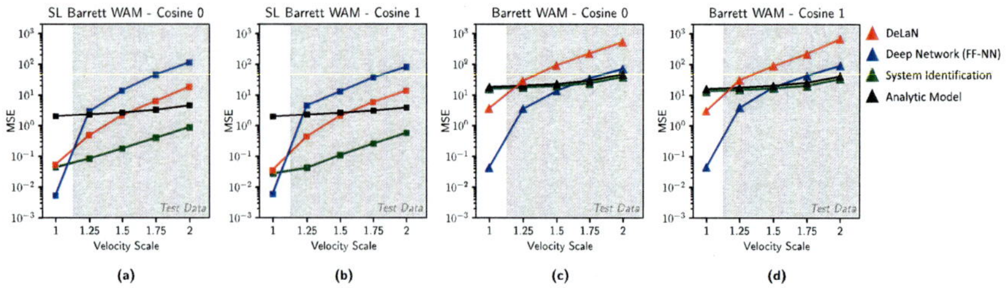

Fig. 3.6 **a**, **b** The mean squared tracking error of the inverse dynamics control following cosine trajectories for the simulated Barrett WAM. **c**, **d** The mean squared tracking error on the physical Barrett WAM. The system identification approach, feed-forward neural network, and DeLaN are trained offline using only the trajectories at a velocity scale of $1\times$. Afterward, the models are tested on the same trajectories with increased velocities to evaluate the extrapolation to new velocities

with the position and derivative gains $\boldsymbol{K}_p$ and $\boldsymbol{K}_d$. In addition, we test the generalization of the learned models by increasing the velocity of the test trajectories. One would expect that DeLaN would generalize better to scaled velocities as the predicted mass matrix and potential energy only depend on the joint position and are independent of the velocity. The training and testing sequences consist of cosine trajectories with different frequencies for each joint and include a little chirp to avoid learning the Fourier basis. The test and train data differ in the frequency of each joint. These trajectories are the standard approach to excite the system and cover a large state domain. The analytic model of the Barrett WAM is obtained from the [52].

The results for the simulated and physical Barrett WAM are summarized in Fig. 3.6. In the simulation, DeLaN and the system identification perform equally well on the training velocity. When comparing generalization, the system identification approach generalizes better than DeLaN to higher velocities. This behavior is expected as system identification obtains the global system parameter while DeLaN only learns a local approximation of the mass matrix and potential energy. In comparison to the feed-forward deep network, DeLaN performs worse on the training velocity but generalizes better to higher velocities. Therefore, the deep network overfits to the training velocity. The analytic model and the system identification have a large performance gap in simulation as we use the same analytic model for simulation and the physical system but the analytic model is optimized for the physical system.

On the physical system, the feed-forward network performs the best on the training domain but deteriorates when the velocity is increased. DeLaN performs worse than the deep network but better than the analytic model and the system identification. The analytic model and the system identification model perform nearly identical. The system identification approach is only marginally better. Both approaches generalize better compared to DeLaN and the deep network. This better generalization is expected as the system parameters are global while the other approaches use local approximations. When increasing the velocity, DeLaN and deep network dynamics model degrade in performance. In contrast to the simulation results, where DeLaN extrapolates better than the deep network, the black-box deep network obtains the better generalization on the physical system. The worse performance and generalization of DeLaN on this physical system can be explained by the assumption of rigid body dynamics. This assumption is not fulfilled due to the cable drives and the motor-side sensing. Therefore, DeLaN cannot model every phenomenon with high fidelity. However, DeLaN learns a good approximation that is better than the system identification approach with the same rigid body assumption.

3.5.3.2 Energy Control Control

A different approach to test the control performance of the learned models is to apply the learned models to controlling under-actuated systems using an energy controller. More specifically we apply an energy controller to swing up the Furuta pendulum and the cartpole. This energy controller regulates the system energy rather than the position and velocities. The control law is described by

$$\boldsymbol{u} = k_E \big[E(\boldsymbol{q}, \dot{\boldsymbol{q}}) - E(\boldsymbol{q}_{\text{des}}, \dot{\boldsymbol{q}}_{\text{des}}) \big]\, sign\big(\dot{\boldsymbol{q}}_1 \cos(\boldsymbol{q}_1)\big) - k_p \boldsymbol{q}_1,$$

with the energy gain k_E and position gain k_p. We use an additional position controller to prevent the system from hitting the joint limits. The control gains are tuned w.r.t. to the analytic model and fixed for all models. This control task is challenging as the control law relies on the system energy, which cannot be learned supervised. Therefore, the feed-forward network baseline cannot be applied to this task. In contrast to the feed-forward network, the physics-inspired deep network models are the first network models that can be applied to this task as these models can infer the system energy. For the training dataset, we use the energy controller to swing-up the pendulum, stabilize the pendulum at the top and let the pendulum fall down after 2 s. Once the pendulum settles the process repeated until about 200 s of data is collected.

The results for the simulated and physical experiments are summarized in Fig. 3.7. Videos of the physical experiments are available at [Link]. Within the simulation, the analytic model, the system identification model, and DeLaN achieve the successful swing-up of the cartpole and Furuta pendulum. On the physical cartpole, all approaches achieve the swing-up despite the large stiction of the linear actuator. For the physical Furuta pendulum, only the analytic model and DeLaN achieve the swing-up. The system identification model does not. The system identification model fails as the linear regression is very sensitive to the observation noise and the small condition number of the features $\boldsymbol{A}$ due to the small dimensions. Therefore, minor changes in the observation can lead to vastly different system parameters. In this specific case, the system identification approach underestimates the masses and hence, exerts too little action to swing up the pendulum and is stuck on the limit cycle.

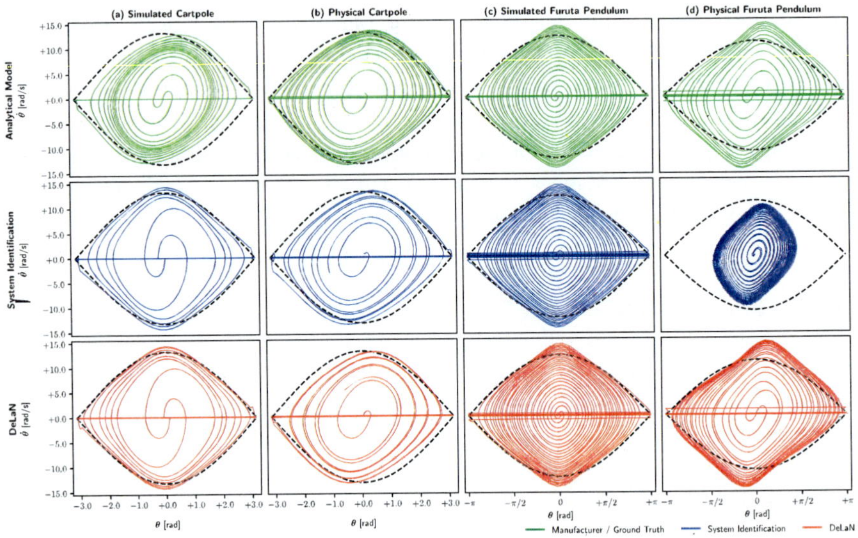

Fig. 3.7 The position θ and velocity $\dot{\theta}$ orbits recorded using energy control to swing up the cartpole and Furuta pendulum. The rows show the different models, i.e., the analytic model, the system identification model, and the DeLaN model while the columns show the different simulated and physical systems. The dashed orbit highlights the desired energy E^*. While the learned and the analytic model can swing up the simulated system and physical Cartpole only the analytic model and DeLaN can swing up the physical Furuta pendulum, while the energy controller using the system identification model cannot

3.5.3.3 Conclusion

The non-linear control experiments on the physical systems show that DeLaN with a structured Lagrangian can learn a good model despite the noisy observations. The resulting model can be used for closed-loop feedback control in real-time for fully-actuated and under-actuated systems. For both systems categories DeLaN achieves a good control performance. It is noteworthy that DeLaN is the first model learning approach utilizing no prior knowledge of the equations of motion that can be applied to energy control. The previous black-box model learning approach could not be applied as the system energy can only be learned unsupervised.

3.6 Conclusion

Coming back to the initial questions of the experiments. The experimental results have showed that

Q1: physics-inspired networks can learn the underlying representation of the dynamical system unsupervised. The predicted inertial, centrifugal, Coriolis and gravitational forces match the ground-truth (Fig. 3.4). On the physical systems the learned system energy can be used for energy control. The swing-up of under-actuated Furuta pendulum and cartpole where successfully achieved.

Q2: physics-inspired networks may learn better models than continuous-time black-box models with feed-forward networks. Especially, in simulation the physics-inspired networks achieve lower mean squared error (Fig. 3.5), longer valid prediction times (Table 3.2) and better generalization (Fig. 3.6). On the physical system, when the assumptions of the physics-inspired networks are violated, the results are not as clear. While DeLaN performs comparable to system identification techniques, the deep network model can represent physical phenomena that are not captured in the physics prior and achieves a lower tracking error.

Q3: physics-inspired networks can be applied to physical systems. Even though the physical systems violate the physics prior due to non-ideal torque sources and unmodeled phenomena such as cable drives, motor dynamics and backslash, DeLaN was able to learn the system energy. For example, the energy controller utilizing DeLaN was able to solve the swing-up the Furuta pendulum and the cartpole, which has large backslash in the linear actuator.

Similar empirical results were also presented by [21, 33–35, 86, 106, 107]. When comparing the different physics prior, Hamiltonian and Lagrangian priors yield comparable models when the corresponding phase space coordinates are observed, i.e., velocities for DeLaN and impulse for HNN. When comparing structured DeLaN and HNN to the black-box DeLaN and HNN, we find that the structured approaches achieve better dynamics models (Table 3.2). The black-box variants struggle to obtain non-singular network Hessians for all possible system states. For the structured approaches singular Hessians can be prevented by parametrized the kinetic energy such that the eigenvalues of the Hessian are lower-bounded and do not become singular.

Despite the advantages of the physics-inspired models to standard deep networks models, the physics-inspired approaches have drawbacks that prevent the general applicability compared to feed-forward networks. In the following, we discuss these limitations.

3.6.1 Open Challenges

Physics-inspired deep networks have two main shortcomings, which have not been solved yet. First of all, the current approaches are only able to simulate articulated rigid-bodies *without* contact and second the current approaches rely on knowing and observing the generalized coordinates. Therefore, most of the existing work only showcased these networks for simple n-link pendulums [108] and n-body problems [87]. For most real-world robotic tasks these assumptions are not full-filled. One frequently does not know or observe the system state and most interesting robotic tasks include contacts. In contrast to physics-inspired networks, black-box dynamics models work with any observations and contacts. These models have been extensively used for model predictive control and are sufficient for complex control tasks [38, 39, 66]. Therefore, these challenges need to be addressed to enable the widespread use of physics-inspired methods for robot control. In the following, we highlight the challenges of both limitation and the initial step towards applying these models to contact-rich tasks with arbitrary observations.

3.6.1.1 Contacts

Analytically, contact forces can be incorporated by adding generalized contact forces to the Euler–Lagrange equation. In this case the differential equation is described by

$$\frac{d}{dt}\frac{\partial \mathcal{L}(\boldsymbol{q},\dot{\boldsymbol{q}})}{\partial \dot{\boldsymbol{q}}} - \frac{\partial \mathcal{L}(\boldsymbol{q},\dot{\boldsymbol{q}})}{\partial \boldsymbol{q}} = \boldsymbol{\tau} + \underbrace{\sum_{i \in \Omega} \boldsymbol{J}_i^c(\boldsymbol{q}) \boldsymbol{f}_i^c(\boldsymbol{q},\dot{\boldsymbol{q}})}_{\text{Generalized Contact Force}},$$

with the Cartesian contact forces $\boldsymbol{f}^c$, the contact Jacobian $\boldsymbol{J}^c$ connecting the Cartesian forces at the contact point to the generalized coordinates and the set of all active contacts Ω. To compute the generalized contact force, analytic simulators first use the known kinematics and meshes to find all contact points and there respective Jacobians. Afterwards the contact force is computed by solving the linear complementarity problem (LCP). A similar approach can also be used for Hamiltonian mechanics that uses contact impulses rather than forces. Within the physics-inspired deep network literature only [46, 108] have included contacts. However, both existing works only consider special cases with strong assumptions. For example, [46] only considers elastic collisions of simple geometric shapes, i.e., circles. In this case, the contact forces can be computed and the contact Jacobian is the identity matrix. Therefore, one only needs to learn an indicator function $\mathbb{1}(\boldsymbol{q})$ being 1 if the contact is active and 0 otherwise. Furthermore, the indicator function is learned supervised. Hence, the training data has to include whether the contact was active or not for each sample. The experiments only apply the proposed algorithm to a ball bouncing on a plane and the Newton cradle.

A different approach was proposed by [108]. This work augments the physics-inspired network with a differentiable physics simulator to handle the contacts. In this case, a collision detection algorithm determines all active contacts and the contact Jacobians. The contact forces are computed by solving the LCP. In this case, only the coefficients of the contact model, e.g., friction and restitution, are learned from data. Therefore, this approach is similar to the white-box friction models described in Sect. 3.4.3.4. This approach also implicitly assumes that the meshes and kinematics are *known*. Without the kinematics and meshes the collision detection algorithms cannot compute the active contacts and Jacobians. If these quantities of the system are known, the analytic equations of motions can be computed and many physical parameters can be approximated from the meshes. Therefore, these assumptions are identical to the required knowledge for system identification using differentiable physics simulators [22, 43, 103]. The advantage of physics-inspired networks compared to system identification with differentiable simulators are unknown. The experiments only applied the proposed approach to bouncing disks and a multi-link pendulum with a ground plane.

To summarize, no general way to add contacts to physics-inspired networks has been proposed and shown to work for multi-contact physics with complex geometries. The naive approach to add a single network to model the generalized contact forces is challenging as this reduces the physics-inspired model learning approaches to a black-box model learning technique without proper regularization. Therefore, an important open challenge for physics-inspired networks for robotics is to introduce a generic approach to include multiple contacts.

3.6.1.2 Generalized Coordinates

The second limiting assumption is the observation of the generalized coordinates $\boldsymbol{q}$, $\dot{\boldsymbol{q}}$ or the generalized momentum $\boldsymbol{p}$. For most robotic systems that do not only involve a rigid body manipulator, these coordinates are commonly not observed or known. One usually only obtains observations derived from the generalized coordinates if the system is fully observed. In many cases, the system is only partially observed and one cannot infer the system state from a single observation. For black-box models this is not a problem as these models do not require specific observations and have been shown to learn good dynamics models for complex systems using only images, e.g., [38–40] and many others.

To overcome this limitation, existing work combined physics-inspired networks with variational autoencoders (VAE) to learn a latent space the resembles the generalized coordinates. In this case, the Lagrangian and Hamiltonian inspired networks are applied in the latent space. Using this approach, the dynamics of single-link pendulums and N-body problems have been learned from artificial images [3, 33, 86, 95, 106]. However, these approaches have not been demonstrated on more complex systems and realistic rendering of systems. Reference [10] also showed that this approach does not necessarily obtain better results than using a normal deep network continuous-time model within the latent space. Therefore, it remains an important open challenge to extend physics-inspired networks to arbitrary observations. The main challenge is to learn a latent space that resembles the generalized coordinates and the naive approach to use a VAE does not seem to be sufficient.

3.6.2 Summary

We introduced physics-inspired networks that combine Lagrangian and Hamiltonian mechanics with deep networks. This combination obtains physically plausible dynamics models that guarantee to conserve energy. The resulting models are also interpretable and can be used as forward, inverse, or energy models using the same parameters. Previously this was not possible with standard deep network dynamics models. Furthermore, we presented all the existing extensions of physics-inspired networks which include different representations of the Hamiltonian and Lagrangian, different loss functions as well as different actuation and friction models. We elaborated on the shortcomings of the current approaches as these techniques are limited to mechanical systems without contacts and require the observation of generalized positions, velocity, momentum, and forces. Therefore, this summary provides the big picture of physics-inspired networks for learning continuous-time dynamics models of rigid body systems.

Within the experiments, we showed that Deep Lagrangian Networks (DeLaN) and Hamiltonian Neural Networks (HNN) learn the underlying structure of the dynamical system for simulated and physical systems. When the corresponding phase-space coordinates of each model are observed, both models perform nearly identical. On average the structured Hamiltonian and Lagrangian perform better than their black-box counterparts. Especially for the Lagrangian combination, the black-box approach can lead to high prediction errors due to inverting the Hessian of a deep network. Furthermore, we show that these physics-inspired techniques can be applied to the physical system despite the observation noise. The resulting DeLaN models can be used for real-time control and achieve good performance for inverse dynamics control as well as energy control. Especially the latter is noteworthy, as DeLaN is the first model learning technique that utilizes deep networks and can learn the system energy. Previously this was only possible using system identification which requires knowledge of the kinematic structure to derive the equations of motion.

References

1. Akkaya, I., Andrychowicz, M., Chociej, M., Litwin, M., McGrew, B., Petron, A., Paino, A., Plappert, M., Powell, G., Ribas, R., et al.: Solving rubik's cube with a robot hand (2019). arXiv:1910.07113
2. Albu-Schäffer, A.: Regelung von Robotern mit elastischen Gelenken am Beispiel der DLR-Leichtbauarme. Ph.D. thesis, Technische Universität München (2002)
3. Allen-Blanchette, C., Veer, S., Majumdar, A., Leonard, N.E.: Lagnetvip: a Lagrangian neural network for video prediction (2020). arXiv:2010.12932
4. Allevato, A., Short, E.S., Pryor, M., Thomaz, A.: Tunenet: One-shot residual tuning for system identification and sim-to-real robot task transfer. In: Conference on Robot Learning (CoRL) (2020)
5. Anderson, B., Hy, T.S., Kondor, R.: Cormorant: covariant molecular neural networks (2019). arXiv:1906.04015
6. Atkeson, C.G., An, C.H., Hollerbach, J.M.: Estimation of inertial parameters of manipulator loads and links. Int. J. Robot. Res. (1986)
7. Bekkers, E.J.: B-spline CNNs on Lie groups. In: International Conference on Learning Representations (ICLR) (2019)
8. Bona, B., Indri, M.: Friction compensation in robotics: an overview. In: 44th IEEE Conference on Decision and Control, 2005 and 2005 European Control Conference. CDC-ECC'05. IEEE (2005)

9. Book, W.J.: Recursive lagrangian dynamics of flexible manipulator arms. Int. J. Robot. Res. (1984)
10. Botev, A., Jaegle, A., Wirnsberger, P., Hennes, D., Higgins, I.: Which priors matter? benchmarking models for learning latent dynamics (2021). arXiv:2111.05458
11. Calinon, S., D'halluin, F., Sauser, E.L., Caldwell, D.G., Billard, A.G.: Learning and reproduction of gestures by imitation. IEEE Robot. & Autom. Mag. (2010)
12. Camacho, E.F., Alba, C.B.: Model Predictive Control. Springer Science & Business Media (2013)
13. Camoriano, R., Traversaro, S., Rosasco, L., Metta, G., Nori, F.: Incremental semiparametric inverse dynamics learning. In: International Conference on Robotics and Automation (ICRA) (2016)
14. Chen, R.T., Rubanova, Y., Bettencourt, J., Duvenaud, D.: Neural ordinary differential equations (2018). arXiv:1806.07366
15. Chen, Z., Zhang, J., Arjovsky, M., Bottou, L.: Symplectic recurrent neural networks (2019). arXiv:1909.13334
16. Choi, Y., Cheong, S.Y., Schweighofer, N.: Local online support vector regression for learning control. In: International Symposium on Computational Intelligence in Robotics and Automation. IEEE (2007)
17. Chu, M., Thuerey, N., Seidel, H.P., Theobalt, C., Zayer, R.: Learning meaningful controls for fluids. ACM Trans. Graph. (TOG) (2021)
18. Cohen, T., Weiler, M., Kicanaoglu, B., Welling, M.: Gauge equivariant convolutional networks and the icosahedral CNN. In: International Conference on Machine Learning (ICML) (2019)
19. Cohen, T., Welling, M.: Group equivariant convolutional networks. In: International Conference on Machine Learning (ICML) (2016)
20. Coumans, E., Bai, Y.: Pybullet, a python module for physics simulation for games, robotics and machine learning. http://pybullet.org (2016)
21. Cranmer, M., Greydanus, S., Hoyer, S., Battaglia, P., Spergel, D., Ho, S.: Lagrangian neural networks (2020). arXiv:2003.04630
22. Degrave, J., Hermans, M., Dambre, J., et al.: A differentiable physics engine for deep learning in robotics. Front. Neurorobot (2019)
23. Dheeru, D., Karra Taniskidou, E.: UCI machine learning repository (2017). http://archive.ics.uci.edu/ml
24. Featherstone, R.: Rigid Body Dynamics Algorithms. Springer (2007)
25. Ferreira, J.P., Crisostomo, M., Coimbra, A.P., Ribeiro, B.: Simulation control of a biped robot with support vector regression. In: IEEE International Symposium on Intelligent Signal Processing. IEEE (2007)
26. Finzi, M., Wang, K.A., Wilson, A.G.: Simplifying hamiltonian and lagrangian neural networks via explicit constraints. In: Advances of Neural Information Processing Systems (NeuRIPS) (2020)
27. Fitzpatrick, R.: Newtonian Dynamics: An Introduction. CRC Press (2021)
28. Gautier, M.: Identification of robots dynamics. IFAC Proc. Vol. (1986)
29. Geist, A.R., Trimpe, S.: Structured learning of rigid-body dynamics: A survey and unified view from a robotics perspective (2020). arXiv:2012.06250
30. Geng, Z., Haynes, L.S., Lee, J.D., Carroll, R.L.: On the dynamic model and kinematic analysis of a class of stewart platforms. Robot. Auton. Syst. (1992)
31. Golliday, C.L., Hemami, H.: An approach to analyzing biped locomotion dynamics and designing robot locomotion controls. IEEE Trans. Autom. Control (1977). https://doi.org/10.1109/TAC.1977.1101650
32. Greenwood, D.T.: Advanced dynamics. Cambridge University Press (2006)
33. Greydanus, S., Dzamba, M., Yosinski, J.: Hamiltonian neural networks. In: Advances in Neural Information Processing Systems (NeuRIPS) (2019)
34. Gupta, J.K., Menda, K., Manchester, Z., Kochenderfer, M.: Structured mechanical models for robot learning and control. In: Learning for Dynamics and Control (2020)

35. Gupta, J.K., Menda, K., Manchester, Z., Kochenderfer, M.J.: A general framework for structured learning of mechanical systems (2019). arXiv:1902.08705
36. Ha, D., Schmidhuber, J.: World models (2018). arXiv:1803.10122
37. Haarnoja, T., Zhou, A., Abbeel, P., Levine, S.: Soft actor-critic: Off-policy maximum entropy deep reinforcement learning with a stochastic actor. In: International Conference on Machine Learning (ICML) (2018)
38. Hafner, D., Lillicrap, T., Ba, J., Norouzi, M.: Dream to control: learning behaviors by latent imagination (2019). arXiv:1912.01603
39. Hafner, D., Lillicrap, T., Fischer, I., Villegas, R., Ha, D., Lee, H., Davidson, J.: Learning latent dynamics for planning from pixels. In: International Conference on Machine Learning (ICML) (2019)
40. Hafner, D., Lillicrap, T., Norouzi, M., Ba, J.: Mastering atari with discrete world models (2020). arXiv:2010.02193
41. Haruno, M., Wolpert, D.M., Kawato, M.: Mosaic model for sensorimotor learning and control. Neural Comput. (2001)
42. Heess, N., Sriram, S., Lemmon, J., Merel, J., Wayne, G., Tassa, Y., Erez, T., Wang, Z., Eslami, S., Riedmiller, M., et al.: Emergence of locomotion behaviours in rich environments (2017). arXiv:1707.02286
43. Heiden, E., Macklin, M., Narang, Y., Fox, D., Garg, A., Ramos, F.: Disect: a differentiable simulation engine for autonomous robotic cutting (2021). arXiv:2105.12244
44. Hemami, H., Wyman, B.: Modeling and control of constrained dynamic systems with application to biped locomotion in the frontal plane. IEEE Trans. Autom. Control (1979). https://doi.org/10.1109/TAC.1979.1102105
45. Hobbs, B.F., Hepenstal, A.: Is optimization optimistically biased? Water Resour. Res. (1989)
46. Hochlehnert, A., Terenin, A., Sæmundsson, S., Deisenroth, M.: Learning contact dynamics using physically structured neural networks. In: International Conference on Artificial Intelligence and Statistics (Aistats) (2021)
47. Holl, P., Koltun, V., Thuerey, N.: Learning to control pdes with differentiable physics (2020). arXiv:2001.07457
48. Huh, I., Yang, E., Hwang, S.J., Shin, J.: Time-reversal symmetric ode network (2020). arXiv:2007.11362
49. Hwangbo, J., Lee, J., Dosovitskiy, A., Bellicoso, D., Tsounis, V., Koltun, V., Hutter, M.: Learning agile and dynamic motor skills for legged robots. Sci. Robot. (2019)
50. Ioannou, P.A., Sun, J.: Robust Adaptive Control. Prentice-Hall (1996)
51. Jansen, M.: Learning an accurate neural model of the dynamics of a typical industrial robot. In: International Conference on Artificial Neural Networks (1994)
52. JHU LCSR, J.L.: Barrett model containing the 7-dof urdf (2018). https://github.com/jhu-lcsr/barrett_model
53. Khansari-Zadeh, S.M., Billard, A.: Learning stable nonlinear dynamical systems with gaussian mixture models. IEEE Trans. Robot. (2011)
54. Khosla, P.K., Kanade, T.: Parameter identification of robot dynamics. In: Conference on Decision and Control (CDC) (1985)
55. Kocijan, J., Murray-Smith, R., Rasmussen, C.E., Girard, A.: Gaussian process model based predictive control. In: American Control Conference (ACC). IEEE (2004)
56. Lagaris, I.E., Likas, A., Fotiadis, D.I.: Artificial neural networks for solving ordinary and partial differential equations. IEEE Trans. Neural Netw. (1998)
57. Lagaris, I.E., Likas, A.C., Papageorgiou, D.G.: Neural-network methods for boundary value problems with irregular boundaries. IEEE Trans. Neural Netw. (2000)
58. Lambert, N., Amos, B., Yadan, O., Calandra, R.: Objective mismatch in model-based reinforcement learning (2020). arXiv:2002.04523
59. Ledezma, F.D., Haddadin, S.: First-order-principles-based constructive network topologies: an application to robot inverse dynamics. In: International Conference on Humanoid Robotics (Humanoids). IEEE (2017)

60. Lee, H., Kang, I.S.: Neural algorithm for solving differential equations. J. Comput. Phys. (1990)
61. Lee, J., Liu, C.K., Park, F.C., Srinivasa, S.S.: A linear-time variational integrator for multibody systems. In: Algorithmic Foundations of Robotics XII. Springer (2020)
62. Lenc, K., Vedaldi, A.: Understanding image representations by measuring their equivariance and equivalence. In: Conference on Computer Vision and Pattern Recognition (2015)
63. Lenz, I., Knepper, R.A., Saxena, A.: Deepmpc: Learning deep latent features for model predictive control. In: Robotics: Science and Systems (2015)
64. Liu, K., Lewis, F., Lebret, G., Taylor, D.: The singularities and dynamics of a stewart platform manipulator. J. Intell. Robot. Syst. (1993)
65. Lurie, A.I.: Analytical Mechanics. Springer Science & Business Media (2013)
66. Lutter, M., Hasenclever, L., Byravan, A., Dulac-Arnold, G., Trochim, P., Heess, N., Merel, J., Tassa, Y.: Learning dynamics models for model predictive agents (2021). arXiv:2109.14311
67. Lutter, M., Peters, J.: Deep lagrangian networks for end-to-end learning of energy-based control for under-actuated systems. In: International Conference on Intelligent Robots and Systems (IROS) (2019)
68. Lutter, M., Ritter, C., Peters, J.: Deep lagrangian networks: Using physics as model prior for deep learning. In: International Conference on Learning Representations (ICLR) (2019)
69. Lutter, M., Silberbauer, J., Watson, J., Peters, J.: A differentiable newton euler algorithm for multi-body model learning (2020). arXiv:2010.09802
70. Lutter, M., Silberbauer, J., Watson, J., Peters, J.: Differentiable physics models for real-world offline model-based reinforcement learning. In: International Conference on Robotics and Automation (ICRA) (2021)
71. Meade, A.J., Jr., Fernandez, A.A.: Solution of nonlinear ordinary differential equations by feedforward neural networks. Math. Comput. Modell. (1994)
72. Miller, K.: The lagrange-based model of delta-4 robot dynamics. Robotersysteme (1992)
73. Mukerjee, A., Ballard, D.: Self-calibration in robot manipulators. In: International Conference on Robotics and Automation (ICRA) (1985)
74. Nguyen-Tuong, D., Peters, J.: Using model knowledge for learning inverse dynamics. In: International Conference on Robotics and Automation (2010)
75. Nguyen-Tuong, D., Seeger, M., Peters, J.: Computed torque control with nonparametric regression models. In: American Control Conference (ACC). IEEE (2008)
76. Nguyen-Tuong, D., Seeger, M., Peters, J.: Model learning with local gaussian process regression. Adv. Robot. (2009)
77. Olsson, H., Åström, K.J., De Wit, C.C., Gäfvert, M., Lischinsky, P.: Friction models and friction compensation. Eur. J. Control (1998)
78. Qin, H.: Machine learning and serving of discrete field theories. Sci. Rep. (2020)
79. Quanser: Quanser courseware and resources (2018). https://www.quanser.com/solution/control-systems/
80. Raissi, M., Perdikaris, P., Karniadakis, G.E.: Physics informed deep learning (part i): Data-driven solutions of nonlinear partial differential equations (2017). arXiv:1711.10561
81. Raissi, M., Perdikaris, P., Karniadakis, G.E.: Physics informed deep learning (part ii): data-driven discovery of nonlinear partial differential equations (2017). arXiv:1711.10566
82. Ratliff, N., Meier, F., Kappler, D., Schaal, S.: Doomed: Direct online optimization of modeling errors in dynamics. Big Data (2016)
83. Romeres, D., Zorzi, M., Camoriano, R., Chiuso, A.: Online semi-parametric learning for inverse dynamics modeling. In: Conference on Decision and Control (CDC) (2016)
84. Romeres, D., Zorzi, M., Camoriano, R., Traversaro, S., Chiuso, A.: Derivative-free online learning of inverse dynamics models. IEEE Trans. Control Syst. Technol. (2019)
85. Rueckert, E., Nakatenus, M., Tosatto, S., Peters, J.: Learning inverse dynamics models in o (n) time with lstm networks. In: International Conference on Humanoid Robotics (Humanoids). IEEE (2017)
86. Saemundsson, S., Terenin, A., Hofmann, K., Deisenroth, M.: Variational integrator networks for physically structured embeddings. In: International Conference on Artificial Intelligence and Statistics (Aistats) (2020)

87. Sanchez-Gonzalez, A., Bapst, V., Cranmer, K., Battaglia, P.: Hamiltonian graph networks with ode integrators (2019). arXiv:1909.12790
88. Sanchez-Gonzalez, A., Heess, N., Springenberg, J.T., Merel, J., Riedmiller, M., Hadsell, R., Battaglia, P.: Graph networks as learnable physics engines for inference and control (2018). arXiv:1806.01242
89. Schaal, S., Atkeson, C.G., Vijayakumar, S.: Scalable techniques from nonparametric statistics for real time robot learning. Appl. Intell. (2002)
90. Schütt, K.T., Arbabzadah, F., Chmiela, S., Müller, K.R., Tkatchenko, A.: Quantum-chemical insights from deep tensor neural networks. Nat. Commun. (2017)
91. Siciliano, B., Khatib, O.: Springer Handbook of Robotics. Springer (2016)
92. Spong, M.W.: Modeling and control of elastic joint robots. J. Dyn. Syst. Meas. Control (1987)
93. Sutanto, G., Wang, A., Lin, Y., Mukadam, M., Sukhatme, G., Rai, A., Meier, F.: Encoding physical constraints in differentiable Newton-Euler algorithm. In: Learning for Dynamics Control (L4DC) (2020)
94. Ting, J.A., Mistry, M., Peters, J., Schaal, S., Nakanishi, J.: A bayesian approach to nonlinear parameter identification for rigid body dynamics. In: Robotics: Science and Systems (2006)
95. Toth, P., Rezende, D.J., Jaegle, A., Racanière, S., Botev, A., Higgins, I.: Hamiltonian generative networks. In: International Conference on Learning Representations (ICLR) (2019)
96. Traversaro, S., Brossette, S., Escande, A., Nori, F.: Identification of fully physical consistent inertial parameters using optimization on manifolds. In: International Conference on Intelligent Robots and Systems (IROS) (2016)
97. Wahrburg, A., Bös, J., Listmann, K.D., Dai, F., Matthias, B., Ding, H.: Motor-current-based estimation of cartesian contact forces and torques for robotic manipulators and its application to force control. IEEE Trans. Autom. Sci. Eng. (2018)
98. Wang, R., Kashinath, K., Mustafa, M., Albert, A., Yu, R.: Towards physics-informed deep learning for turbulent flow prediction. In: International Conference on Knowledge Discovery & Data Mining (2020)
99. Wang, R., Walters, R., Yu, R.: Incorporating symmetry into deep dynamics models for improved generalization (2020). arXiv:2002.03061
100. Weiler, M., Cesa, G.: General $e(2)$-equivariant steerable CNNs (2019). arXiv:1911.08251
101. Wells, D.A.: Schaum's Outline of Theory and Problems of Lagrangian Dynamics. McGraw-Hill (1967)
102. Wensing, P.M., Kim, S., Slotine, J.J.E.: Linear matrix inequalities for physically consistent inertial parameter identification: a statistical perspective on the mass distribution. IEEE Robot. Autom. Lett. (RAL) (2017)
103. Werling, K., Omens, D., Lee, J., Exarchos, I., Liu, C.K.: Fast and feature-complete differentiable physics for articulated rigid bodies with contact (2021). arXiv:2103.16021
104. Williams, B., Toussaint, M., Storkey, A.J.: Modelling motion primitives and their timing in biologically executed movements. In: Advances in Neural Information Processing Systems (NeurIPS) (2008)
105. de Wit, C.C., Siciliano, B., Bastin, G.: Theory of Robot Control. Springer Science & Business Media (2012)
106. Zhong, Y.D., Dey, B., Chakraborty, A.: Symplectic ode-net: learning hamiltonian dynamics with control (2019). arXiv:1909.12077
107. Zhong, Y.D., Dey, B., Chakraborty, A.: Dissipative symoden: encoding hamiltonian dynamics with dissipation and control into deep learning (2020). arXiv:2002.08860
108. Zhong, Y.D., Dey, B., Chakraborty, A.: A differentiable contact model to extend Lagrangian and Hamiltonian neural networks for modeling hybrid dynamics (2021). arXiv:2102.06794
109. Zhou, K., Doyle, J.C., Glover, K., et al.: Robust and Optimal Control. Prentice Hall (1996)

Chapter 4
Continuous-Time Fitted Value Iteration for Robust Policies

4.1 Introduction

One approach to obtain the optimal control inputs that maximize the reward, is to solve the Hamilton–Jacobi–Bellman (HJB) equation, as this differential equation expresses a sufficient and necessary condition for optimality [45]. Solving the HJB yields the optimal value function, which can be used to retrieve the optimal action at each state. Therefore, this ansatz has been used by various research communities, including economics [13, 14] and robotics [4, 17, 25, 44], to compute the optimal plan for a given reward function. For example in robotics, the optimal action sequence to navigate a robot to a goal from any starting state with the least actions can be obtained by solving the HJB. Classical approaches solve this differential equation using PDE solvers on a discretized grid [4, 17, 25]. Instead of using a grid, various researchers have proposed to use the machine-learning toolset of black-box function approximation and regression techniques to solve the HJB using randomly sampled data [21, 40, 41, 48, 49, 57, 79, 87].

In this chapter, we follow this line of research and present an approach to solve the HJB and the adversarial extension of the HJB using value iteration. This approach unifies the derivation of our previously proposed algorithms Continuous Fitted Value Iteration (cFVI) [51] and Robust Fitted Value Iteration (rFVI) [50]. cFVI is a value iteration based algorithm that solves the HJB for continuous states and actions. This approach leverages our insights to obtain the optimal policy in closed form for control-affine dynamics and separable rewards. This analytic policy enables us to solve this differential equation using fitted value iteration with a deep network as value function approximation. Previously this would not have been possible for continuous actions as solving for the optimal action at each step would not be computationally feasible. rFVI is similar to cFVI but instead of solving the HJB, this algorithms solves the Hamilton–Jacobi–Isaacs (HJI). The HJI incorporates an additional adversary that tries to minimize the reward. Therefore, the obtained policy and value function are

M. Lutter, *Inductive Biases in Machine Learning for Robotics and Control*,
Springer Tracts in Advanced Robotics 156,
https://doi.org/10.1007/978-3-031-37832-4_4

robust to perturbations of the adversary. For this extension, we show that also the optimal perturbation of the adversary can be computed in closed form.

We apply the resulting algorithms to standard continuous control problems. In this setting, the HJI is of special interest as the adversary is used to control the parameters of the environment to minimize the reward [4, 17, 25]. Therefore, this min-max formulation optimizes the worst-case reward. This worst-case optimization yields an optimal policy that is robust to changes in the environment because the worst-case is assumed during planning. We show that the proposed approaches can obtain the optimal control policy and can be transferred to the physical system. By changing the masses of the dynamical systems during the simulation to real (Sim2Real) transfer experiments, we show that the rFVI policy is more robust compared to the baselines.

In this chapter, we show that cFVI and rFVI obtain the optimal control policy and can be successfully transferred to a physical system. To derive these algorithms and highlight their performance,

1. We extend the existing derivations [21, 54, 57] of the optimal policy to a wider class of reward functions and adversaries,
2. We propose to use value iteration to solve the differential equations as this optimization is more robust compared to the approaches using regression [21, 49, 79],
3. We provide an extensive experimental evaluation that compares qualitative and quantitative performance evaluates the policies on the physical system using Sim2Real. Furthermore, we provide ablation studies highlighting the impact of the individual hyper-parameters.

4.1.1 Outline

The chapter is structured as follows. First, we introduce the problem statement (Sect. 4.2). Afterwards, we derive the analytic optimal policy (Sect. 4.3), extend the approach to action constraints (Sect. 4.3.1) and derive the optimal adversary (Sect. 4.3.2). Section 4.4 introduces the value iteration to compute the optimal value function and the following Section describes the used value function representation (Sect. 4.4.4). The experiments are summarized in Sect. 4.5. The following related work section (Sect. 4.6) compares cFVI to the existing literature. Section 4.7 discusses the observed limitations and potential future work. Finally, Sect. 4.8 summarizes the chapter.

4.2 Problem Statement

We focus on solving the Hamilton–Jacobi–Bellman and Hamilton–Jacobi–Isaacs differential equation. These equations can be derived using the continuous-time RL problem and the corresponding adversarial extension. In the following, we first introduce the continuous-time RL problem and extend it to the adversarial formulation afterward.

4.2.1 Reinforcement Learning

The infinite horizon continuous time reinforcement learning problem is described by

$$\pi^*(\boldsymbol{x}_0) = \arg\max_{\boldsymbol{u}} \int_0^{\infty} \exp(-\rho t)\ r_c(\boldsymbol{x}_t, \boldsymbol{u}_t)\, dt, \tag{4.1}$$

$$V^*(\boldsymbol{x}_0) = \max_{\boldsymbol{u}} \int_0^{\infty} \exp(-\rho t)\, r_c(\boldsymbol{x}_t, \boldsymbol{u}_t)\, dt, \tag{4.2}$$

$$\text{with} \quad \boldsymbol{x}(t) = \boldsymbol{x}_0 + \int_0^{t} f_c(\boldsymbol{x}_\tau, \boldsymbol{u}_\tau)\, d\tau \tag{4.3}$$

with the discounting factor $\rho \in (0, \infty]$, the reward r_c and the dynamics f_c [42]. Notably, the discrete-time reward and discounting can be described using the continuous-time counterparts, i.e., $r(\boldsymbol{x}, \boldsymbol{u}) = \Delta t\, r_c(\boldsymbol{x}, \boldsymbol{u})$ and $\gamma = \exp(-\rho\, \Delta t)$ with the sampling interval Δt. The continuous-time discounting ρ is, in contrast to the discrete discounting factor γ, agnostic to sampling frequencies. In the continuous-time case, the Q-function does not exist [21]. It is important to note that the optimization of Eq. 4.2 is unconstrained w.r.t. to the actions. Action constraints will be implicitly introduced using the action cost in Sect. 4.3.1.

Assuming that the optimal value function is differentiable, Eq. 4.2 can be rewritten to yield the HJB differential equation, which is the continuous time counterpart of the discrete Bellmann equation. Substituting the value function at time $t' = t + \Delta t$, approximating $V^*(x(t'),\ t')$ with its 1st order Taylor expansion and taking the limit $\Delta t \to 0$ yields the HJB described by

$$\rho\ V^*(\boldsymbol{x}) = \max_{\boldsymbol{u}}\ r(\boldsymbol{x}, \boldsymbol{u}) + f_c(\boldsymbol{x}, \boldsymbol{u})^\top \frac{\partial V^*}{\partial \boldsymbol{x}}. \tag{4.4}$$

In the following, we will abbreviate $\partial V^* / \partial \boldsymbol{x}$ as $\nabla_x V^*$. The full derivation of the HJB can be found within [21]. The reward is assumed to be separable into a non-linear state reward q_c and the action cost g_c described by

$$r_c(\boldsymbol{x}, \boldsymbol{u}) = q_c(\boldsymbol{x}) - g_c(\boldsymbol{u}). \tag{4.5}$$

The action penalty g_c is non-linear, positive definite, and strictly convex. This separability is common for robot control problems as rewards are composed of a state component quantifying the distance to the desired state and an action penalty. The action cost penalizes non-zero actions to avoid bang-bang control from being optimal and is convex to have a unique optimal action.

The deterministic continuous-time dynamics model f_c is assumed to be non-linear w.r.t. the system state $\boldsymbol{x}$ but affine w.r.t. the action $\boldsymbol{u}$. Such dynamics model is described by

$$\dot{\boldsymbol{x}} = \boldsymbol{a}(\boldsymbol{x}; \theta) + \boldsymbol{B}(\boldsymbol{x}; \theta)\boldsymbol{u} \tag{4.6}$$

with the non-linear drift $\boldsymbol{a}$, the non-linear control matrix $\boldsymbol{B}$ and the system parameters θ. Robot dynamics models are naturally expressed in the continuous-time formulation and many are control affine. Furthermore, this special case has received ample attention in the existing literature due to its wide applicability [21, 38, 83].

4.2.2 *Adversarial Reinforcement Learning*

The adversarial approach incorporates an adversary that controls the environment and tries to minimize the obtained reward of the policy. This formulation resembles an zero-sum two-player game, where the policy maximizes the reward and the adversary minimizes the reward. Therefore, one optimizes the worst case reward and not the expected reward. The worst-case formulation is commonly used within robust control to obtain a policy that is robust to changes in the environments. The corresponding optimization problems of Eqs. 4.1 and 4.2 are described by

$$\pi^*(\boldsymbol{x}) = \arg\max_{\pi} \inf_{\boldsymbol{\xi}\in\Omega} \int_0^\infty \exp(-\rho t)\; r_c(\boldsymbol{x}_t, \boldsymbol{u}_t)\, dt, \tag{4.7}$$

$$V^*(\boldsymbol{x}) = \max_{\boldsymbol{u}} \inf_{\boldsymbol{\xi}\in\Omega} \int_0^\infty \exp(-\rho t)\, r_c(\boldsymbol{x}_t, \boldsymbol{u}_t)\, dt, \tag{4.8}$$

with the adversary $\boldsymbol{\xi}$, admissible set Ω. The order of the optimizations can be switched as the optimal actions and disturbance remain identical [36]. The adversary $\boldsymbol{\xi}$ actions are constrained to be in the set of admissible disturbances Ω as otherwise the adversary is too powerful and would prevent the policy from learning the task. Similar to the HJB, Eq. 4.8 can be rewritten to yield the Hamilton–Jacobi–Isaacs (HJI) differential equation. The HJI is described by

$$\rho\, V^*(\boldsymbol{x}) = \max_{\boldsymbol{u}} \inf_{\boldsymbol{\xi}\in\Omega}\; r(\boldsymbol{x}, \boldsymbol{u}) \;+ f_c(\boldsymbol{x}, \boldsymbol{u}, \boldsymbol{\xi})^\top\, \nabla_{\boldsymbol{x}} V^*. \tag{4.9}$$

The optimal policy and adversary are assumed to be stationary and Markovian. In this case, the worst-case action is deterministic if the dynamics are deterministic. If the adversary would be stochastic, the optimal policies are non-stationary and non-Markovian [88]. This assumption is frequently used in the adversarial RL literature [33, 55, 57, 88].

To obtain a policy that is robust to variations of the dynamical system and bridge the simulation to reality gap, we consider four different adversaries.

These adversaries either alter (1) the state [33, 47, 61], (2) action [27, 57, 63, 66, 67, 81], (3) observation [55, 88] and (4) model parameters [55]. Each adversary addresses a potential cause of the simulation gap. The state adversary $\boldsymbol{\xi}_x$ incorporates unmodeled physical phenomena in the simulation. The action adversary $\boldsymbol{\xi}_u$ addresses the non-ideal actuators. The observation adversary $\boldsymbol{\xi}_o$ introduces the non-ideal observations caused by sensors. The model adversary $\boldsymbol{\xi}_\theta$ introduces a bias to the system parameters. All adversaries could be subsumed via a single adversary with a large admissible set. However, the resulting dynamics would not capture the underlying structure of the simulation gap [55] and the optimal policy would be too conservative [86]. Therefore, we disambiguate between the different adversaries to capture this structure. However, all four adversaries can be combined to obtain robustness against each variation. Mathematically, the system dynamics including the adversary are

$$\text{State } \boldsymbol{\xi}_x: \quad \dot{\boldsymbol{x}} = \boldsymbol{a}(\boldsymbol{x};\boldsymbol{\theta}) + \boldsymbol{B}(\boldsymbol{x};\boldsymbol{\theta})\boldsymbol{u} + \boldsymbol{\xi}_x, \tag{4.10}$$

$$\text{Action } \boldsymbol{\xi}_u: \quad \dot{\boldsymbol{x}} = \boldsymbol{a}(\boldsymbol{x};\boldsymbol{\theta}) + \boldsymbol{B}(\boldsymbol{x};\boldsymbol{\theta})\left(\boldsymbol{u} + \boldsymbol{\xi}_u\right), \tag{4.11}$$

$$\text{Observation } \boldsymbol{\xi}_o: \quad \dot{\boldsymbol{x}} = \boldsymbol{a}(\boldsymbol{x} + \boldsymbol{\xi}_o;\boldsymbol{\theta}) + \boldsymbol{B}(\boldsymbol{x} + \boldsymbol{\xi}_o;\boldsymbol{\theta})\,\boldsymbol{u}, \tag{4.12}$$

$$\text{Model } \boldsymbol{\xi}_\theta: \quad \dot{\boldsymbol{x}} = \boldsymbol{a}(\boldsymbol{x};\boldsymbol{\theta} + \boldsymbol{\xi}_\theta) + \boldsymbol{B}(\boldsymbol{x};\boldsymbol{\theta} + \boldsymbol{\xi}_\theta)\,\boldsymbol{u}. \tag{4.13}$$

Instead of disturbing the observation, Eq. 4.12 disturbs the simulation state of the drift and control matrix. This disturbance is identical to changing the observed state.

4.3 Deriving the Optimal Policy

As a first step to solve the HJB (Eq. 4.4) and the HJI (Eq. 4.9), one must obtain an efficient approach to solve the maximization w.r.t. the actions. In the case of discrete actions, this optimization can be solved by evaluating each action and choosing the action with the highest value. In the continuous action case, one cannot evaluate each action, and numerically solving an optimization problem at each state is computationally too expensive. Therefore, one requires an analytic solution to the optimization. This closed-form solution enables a computationally efficient algorithm to solve both differential equations. In the following, we show that this optimization can be solved using the previously described assumptions.

Theorem 4.1 *If the dynamics are control affine (Eq. 4.6), the reward is separable w.r.t. to state and action (Eq. 4.5) and the action cost g_c is positive definite and strictly convex, the continuous-time optimal policy π^* is described by*

$$\pi^*(\boldsymbol{x}) = \nabla \tilde{g}_c\left(\boldsymbol{B}(\boldsymbol{x})^\top \nabla_x V^*\right) \tag{4.14}$$

where $\tilde{g}$ is the convex conjugate of g and $\nabla_x V^*$ is the Jacobian of current value function V^* w.r.t. the system state.

Proof The detailed proof is provided in the appendix. This derivation follows our previous work [49], which generalized the special cases initially described by Lyshevski [54] and Doya [21] to a wider class of reward functions. Substituting Eqs. 4.6 and 4.5 into the HJB (Eq. 4.4) yields

$$\rho V^*(\boldsymbol{x}) = \max_{\boldsymbol{u}} \; q_c(\boldsymbol{x}) - g_c(\boldsymbol{u}) + \nabla_x V^\top [\boldsymbol{a}(\boldsymbol{x}) + \boldsymbol{B}(\boldsymbol{x})\boldsymbol{u}] \, .$$

Therefore, the optimal action is described by

$$\boldsymbol{u}_t^* = \arg\max_{\boldsymbol{u}} \; \nabla_x V^\top \boldsymbol{B}(\boldsymbol{x}_t) \, \boldsymbol{u} - g_c(\boldsymbol{u}). \tag{4.15}$$

This optimization can be solved analytically as g_c is strictly convex and hence $\nabla g_c(\boldsymbol{u}) = \boldsymbol{w}$ is invertible, i.e., $\boldsymbol{u} = [\nabla g_c]^{-1}(\boldsymbol{w}) = \nabla \tilde{g}_c(\boldsymbol{w})$ with the convex conjugate $\tilde{g}$. The solution of Eq. 4.15 is described by

$$\boldsymbol{B}^\top \nabla_x V^* - \nabla g_c(\boldsymbol{u}) = 0 \;\Rightarrow\; \boldsymbol{u}^* = \nabla \tilde{g}_c \left(\boldsymbol{B}^\top \nabla_x V^* \right) .$$

This closed-form optimal policy has an intuitive interpretation. The policy performs steepest ascent by following the gradient of the value function. The inner part $\boldsymbol{B}(\boldsymbol{x})^\top \nabla_x V^k$ projects the change in state onto the action space. The action cost g_c determines the magnitude of the action. The projected gradient is then reshaped using the action cost. The design of the action cost will be explained in the next section. If the continuous-time optimal policy is executed by a discrete-time controller, the time discretization determines the step-size of the hill climbing. Therefore, the time discretization affects the convergence to the value function maximum. If the step size is too large the system becomes unstable and does not converge to the maximum. For most real-world robotic systems with natural frequencies below 5 Hz and control frequencies above 100 Hz, the resulting step-size is sufficiently small to achieve convergence to the value function maximum. Therefore, π^* can be used for high-frequency discrete-time controllers. Furthermore, the continuous-time policy can be used for intermittent control (event-based control) where interacting with the system occurs at irregular time-steps and each interaction is associated with a cost [2].

4.3.1 Action Constraints

The previous Section derived the analytic expression that describes the impact of the action cost on the optimal actions. Therefore, the action cost can be used to design the shape of the policy. By selecting a specific shape of the policy, one can leverage the convex conjugacy to derive the corresponding action cost. The shape of the optimal policy is determined by the monotone function ∇g^*. Therefore, one can define any desired monotone shape and determine the corresponding strictly convex cost by inverting ∇g^* to compute ∇g and integrating ∇g to obtain the strictly convex cost function $g(\boldsymbol{u})$. This approach can be used to design action costs such that the common standard controllers become optimal. For example, bang-bang control is optimal w.r.t. to no action cost. The linear policy is optimal w.r.t. the quadratic action cost. The logistic policy is optimal w.r.t. the binary cross-entropy cost. The Atan shaped policy is optimal w.r.t. the log cosine cost. Incorporating the action constraints directly via barrier-shaped action cost is beneficial as clipping the unbounded actions is only optimal for linear systems [20] and increasing the quadratic action cost to ensure the action limits leads to over-conservative behavior and underuse of the control range. Furthermore, this implicit integration of the action limits via the action cost enables to only solve the unconstrained optimization problem rather than incorporating the action constraints via an explicit constraint. The full generality of this concept based on convex conjugacy is shown in Table 4.1, which shows the corresponding cost functions for linear, logistic, atan, tanh, and bang-bang controllers.

Using the rules from convex analysis [11], the action limits and action range can be adapted as shown by Action-Scaled, Action-Shifted, and Cost-Scaled rows in Table 4.1. This enables quick experimentation by mixing and matching costs. For example, the action cost corresponding to the tanh policy is straightforwardly derived using the well-known relationship between $\tanh(x)$ and the logistic sigmoid $\sigma(x)$ given by $\tanh(x) = 2\sigma(2x) - 1$. Note that a formula for general invertible affine transformations can be derived, not only for scalar scaling and shifting. Classical types of hard nonlinearities [18] can be derived as limiting cases of smooth solutions. For example, taking the Tanh action cost g_{tanh} and scaling it with $\beta \to 0$, i.e., putting a very small cost on actions, results in the Bang-Bang control shape. Taking a different limit of the Tanh policy in which scaling is performed simultaneously w.r.t. the action and cost, the resulting shape is what we call Bang-Lin and corresponds to a function which is linear around zero and saturates for larger input values.

4.3.2 Optimal Adversary Actions

To solve the HJI efficiently one not only requires the optimal policy to be described using an analytic form but also the optimal adversary. To obtain this solution one must solve the constrained min-max optimization of Eq. 4.9. We show that this optimization problem can be solved analytically for the described dynamics and disturbance models using the Karush–Kuhn–Tucker conditions. It is important to note that the adversaries are not exclusive and can be combined. We only derive the individual cases for simplicity.

The resulting optimal actions $\boldsymbol{u}^*$ and disturbances $\boldsymbol{\xi}_i^*$ have a coherent intuitive interpretation. The optimal actions perform steepest ascent by following the gradient of the value function $\nabla_x V$. The optimal perturbations perform steepest descent by following the negative gradient of the value function. The magnitude of taken action is determined by the action cost g in the case of the optimal policy or the admissible set Ω in the case of the adversary. The optimal policy and the optimal adversary is described by

Table 4.1 Selected action costs. The choice of the action cost $g(\boldsymbol{u})$ determines the range of actions $\boldsymbol{u} \in \mathrm{dom}\,(g)$, as well as the form of the optimal policy $\nabla g^*(\boldsymbol{w})$ and the type of non-linearity in the HJB equation $g^*(\boldsymbol{w})$. Section 1 of the table contains policies with standard action domains. Section 2 provides formulae for shifting and scaling actions and scaling costs. Section 3 & 4 show how to use the formulaes

Policy name	Action range	Action cost $g(\boldsymbol{u})$	Policy $\nabla g^*(\boldsymbol{w})$	HJB nonlinear term $g^*(\boldsymbol{w})$
Linear	$\boldsymbol{u} \in \mathbb{R}^{n_u}$	$\frac{1}{2}\boldsymbol{u}^\top \mathbf{R}\boldsymbol{u}$	$\mathbf{R}^{-1}\boldsymbol{w}$	$\frac{1}{2}\boldsymbol{w}^\top \mathbf{R}^{-1}\boldsymbol{w}$
Logistic	$\mathbf{0} < \boldsymbol{u} < \mathbf{1}$	$\boldsymbol{u}^\top \log \boldsymbol{u} + (\mathbf{1} - \boldsymbol{u})^\top \log(\mathbf{1} - \boldsymbol{u})$	$\frac{\mathbf{1}}{\mathbf{1}+e^{-\boldsymbol{w}}} =: \sigma(\boldsymbol{w})$	$\mathbf{1}^\top \log(\mathbf{1} + e^{\boldsymbol{w}})$
Atan	$-\frac{\pi}{2}\mathbf{1} < \boldsymbol{u} < \frac{\pi}{2}\mathbf{1}$	$-\log\cos \boldsymbol{u}$	$\tan^{-1}(\boldsymbol{w})$	$\boldsymbol{w}^\top \tan^{-1}(\boldsymbol{w}) - \frac{1}{2}\mathbf{1}^\top \log(\mathbf{1} + \boldsymbol{w}^2)$
Action-Scaled	$\boldsymbol{u} \in \alpha\,\mathrm{dom}\,(g)$	$\alpha g(\alpha^{-1}\boldsymbol{u})$	$\alpha \nabla g^*(\boldsymbol{w})$	$\alpha g^*(\boldsymbol{w})$
Cost-Scaled	$\boldsymbol{u} \in \mathrm{dom}\,(g)$	$\beta g(\boldsymbol{u})$	$\nabla g^*(\beta^{-1}\boldsymbol{w})$	$\beta g^*(\beta^{-1}\boldsymbol{w})$
Action-Shifted	$\boldsymbol{u} \in \mathrm{dom}\,(g) - \gamma\mathbf{1}$	$g(\boldsymbol{u} + \gamma\mathbf{1}) - g(\gamma\mathbf{1})$	$\nabla g^*(\boldsymbol{w}) - \gamma\mathbf{1}$	$g^*(\boldsymbol{w}) - \gamma\mathbf{1}^\top \boldsymbol{w}$
Tanh	$-\mathbf{1} < \boldsymbol{u} < \mathbf{1}$	$g_{\text{logistic}}(\frac{\boldsymbol{u}+\mathbf{1}}{2}) - g_{\text{logistic}}(\frac{\mathbf{1}}{2})$	$\tanh \boldsymbol{w} = 2\sigma(2\boldsymbol{w}) - \mathbf{1}$	$\mathbf{1}^\top \log\cosh \boldsymbol{w}$
TanhActScaled	$-\alpha\mathbf{1} < \boldsymbol{u} < \alpha\mathbf{1}$	$\alpha g_{\tanh}(\alpha^{-1}\boldsymbol{u})$	$\alpha \tanh \boldsymbol{w}$	$\alpha\mathbf{1}^\top \log\cosh \boldsymbol{w}$
AtanActScaled	$-\alpha\mathbf{1} < \boldsymbol{u} < \alpha\mathbf{1}$	$-\frac{2\alpha}{\pi}\log\cos(\frac{2\alpha}{\pi}\boldsymbol{u})$	$\frac{2\alpha}{\pi}\tan^{-1}(\boldsymbol{w})$	$\frac{2\alpha}{\pi} g^*_{\text{atan}}(\boldsymbol{w})$
Bang-Bang	$-\mathbf{1} \leq \boldsymbol{u} \leq \mathbf{1}$	$\chi_{[-\mathbf{1},\mathbf{1}]}(\boldsymbol{u})$, χ - charact. fun.	$sign \boldsymbol{w}$	$\Vert\boldsymbol{w}\Vert_1$
Bang-Lin	$-\mathbf{1} \leq \boldsymbol{u} \leq \mathbf{1}$	$\frac{1}{2}\boldsymbol{u}^\top \boldsymbol{u}\ \chi_{[-\mathbf{1},\mathbf{1}]}(\boldsymbol{u})$	$-\mathbf{1} + \sum_{\delta=-1}^{1} \mathrm{relu}(\mathbf{1} - \delta\boldsymbol{w})$	$\mathbf{1}^\top L_1(\boldsymbol{w})$, $L_\delta(a)$—Huber loss

$$\boldsymbol{u}^* = \nabla\tilde{g}\left(\frac{\partial f_c(.)}{\partial \boldsymbol{u}}^\top \nabla_x V^*\right), \qquad \boldsymbol{\xi}_i^* = -h_\Omega\left(\frac{\partial f_c(.)}{\partial \boldsymbol{\xi}_i}^\top \nabla_x V^*\right). \tag{4.16}$$

In the following we abbreviate $\left[\partial f_c(.)/\partial \boldsymbol{y}\right]^\top \nabla_x V$, as z_y. For the adversarial policy, h_Ω rescales z_ξ to be on the boundary of the admissible set. If the admissible set bounds the signal energy to be smaller than α, the disturbance is rescaled to have the length α. Therefore, the adversary is described by

$$\Omega_E = \{\boldsymbol{\xi} \in \mathbb{R}^n \mid \|\boldsymbol{\xi}\|_2 \leq \alpha\} \;\Rightarrow\; h_E(z_\xi) = \alpha \frac{z_\xi}{\|z_\xi\|_2}. \tag{4.17}$$

If the amplitude of the disturbance is bounded, the disturbance performs bang-bang control. In this case the adversarial policy is described by

$$\begin{aligned} \Omega_A = \{\boldsymbol{\xi} \in \mathbb{R}^n \mid \boldsymbol{\nu}_{\min} \leq \boldsymbol{\xi} \leq \boldsymbol{\nu}_{\max}\} \\ \Rightarrow\; h_A(z_\xi) = \boldsymbol{\Delta} sign\left(z_\xi\right) + \boldsymbol{\mu}, \end{aligned} \tag{4.18}$$

with $\boldsymbol{\mu} = (\boldsymbol{\nu}_{\max} + \boldsymbol{\nu}_{\min})/2$ and $\boldsymbol{\Delta} = (\boldsymbol{\nu}_{\max} - \boldsymbol{\nu}_{\min})/2$.

The following theorem derive Eqs. 4.16, 4.17 and 4.18 for the optimal policy and the different disturbances. Following the theorem, we provide sketches of the proofs for the state and model disturbance. The remaining proofs are analogous. The complete proofs for all theorems are provided in the appendix. All solutions are summarized in Table 4.2.

Theorem 4.2 *If the dynamics are control affine (Eq. 4.6), the reward is separable w.r.t. to state and action (Eq. 4.5) and the action cost g_c is positive definite and strictly convex, the continuous-time optimal policy π^* and optimal adversary $\boldsymbol{\xi}^*$ can be computed in closed form.*

2.1 State Disturbance. *The optimal policy π^* and state disturbance $\boldsymbol{\xi}_x$ (Eq. 4.10) with bounded signal energy (Eq. 4.17) is described by*

$$\pi^*(\boldsymbol{x}) = \nabla\tilde{g}\left(\boldsymbol{B}(\boldsymbol{x})^\top \nabla_x V^*\right), \qquad \boldsymbol{\xi}_x^* = -\alpha \frac{\nabla_x V^*}{\|\nabla_x V^*\|_2}.$$

2.2 Action Disturbance. *The optimal policy π^* and action disturbance $\boldsymbol{\xi}_u$ (Eq. 4.11) with bounded signal energy (Eq. 4.17) is described by*

$$\pi^*(\boldsymbol{x}) = \nabla\tilde{g}\left(\boldsymbol{B}(\boldsymbol{x})^\top \nabla_x V^*\right), \qquad \boldsymbol{\xi}_u = -\alpha \frac{\boldsymbol{B}(\boldsymbol{x})^\top \nabla_x V^*}{\|\boldsymbol{B}(\boldsymbol{x})^\top \nabla_x V^*\|_2}.$$

2.3 Observation Disturbance. *The optimal policy π^* and observation disturbance $\boldsymbol{\xi}_\theta$ (Eq. 4.12) with bounded signal energy (Eq. 4.17), smooth drift and control matrix (i.e., $\boldsymbol{a}, \boldsymbol{B} \in C^1$) and $\boldsymbol{B}(\boldsymbol{x} + \boldsymbol{\xi}_o) \approx \boldsymbol{B}(\boldsymbol{x})$ is described by*

Table 4.2 The optimal actions $\boldsymbol{u}^k$ and adversarial actions ξ^k for the state-, action-, model- and observation bias with the admissible set Ω

	State perturbation	Action perturbation	Model perturbation	Observation perturbation
Dynamics	$\boldsymbol{a}(\boldsymbol{x}) + \boldsymbol{B}(\boldsymbol{x})\boldsymbol{u} + \xi$	$\boldsymbol{a}(\boldsymbol{x}) + \boldsymbol{B}(\boldsymbol{x})(\boldsymbol{u} + \xi)$	$\boldsymbol{a}(\theta + \xi) + \boldsymbol{B}(\theta + \xi)\boldsymbol{u}$	$\boldsymbol{a}(\boldsymbol{x} + \xi) + \boldsymbol{B}(\boldsymbol{x} + \xi)\boldsymbol{u}$
Opt. Action $\boldsymbol{u}^k$	$\nabla\tilde{g}(\boldsymbol{B}(\boldsymbol{x})^\top \nabla_x V^k)$	$\nabla\tilde{g}(\boldsymbol{B}(\boldsymbol{x})^\top \nabla_x V^k)$	$\nabla\tilde{g}(\boldsymbol{B}(\boldsymbol{x})^\top \nabla_x V^k)$	$\nabla\tilde{g}(\boldsymbol{B}(\boldsymbol{x})^\top \nabla_x V^k)$
Opt. Disturbance ξ^k	$-h_\Omega\left(\nabla_x V^k\right)$	$-h_\Omega\left(\boldsymbol{B}^\top \nabla_x V^k\right)$	$-h_\Omega\left(\left(\frac{\partial \boldsymbol{a}}{\partial \boldsymbol{x}} + \frac{\partial \boldsymbol{B}}{\partial \theta}\boldsymbol{u}^k\right)^\top \nabla_x V^k\right)$	$-h_\Omega\left(\left(\frac{\partial \boldsymbol{a}}{\partial \boldsymbol{x}} + \frac{\partial \boldsymbol{B}}{\partial \boldsymbol{x}}\boldsymbol{u}^k\right)^\top \nabla_x V^k\right)$

$$\pi(\boldsymbol{x}) = \nabla\tilde{g}\left(\boldsymbol{B}(\boldsymbol{x})^\top \nabla_x V\right), \qquad \boldsymbol{\xi}_o = -\alpha \frac{z_o}{\|z_o\|_2}$$
$$\text{with } z_o = \left(\frac{\partial \boldsymbol{a}(\boldsymbol{x};\ \boldsymbol{\theta})}{\partial \boldsymbol{x}} + \frac{\partial \boldsymbol{B}(\boldsymbol{x};\ \boldsymbol{\theta})}{\partial \boldsymbol{x}} \pi(\boldsymbol{x})\right)^\top \nabla_x V.$$

2.4 Model Disturbance. *The optimal policy π^* and model disturbance $\boldsymbol{\xi}_\theta$ (Eq. 4.13) with element-wise bounded amplitude (Eq. 4.18), smooth drift and control matrix (i.e., $\boldsymbol{a}, \boldsymbol{B} \in C^1$) and $\boldsymbol{B}(\boldsymbol{\theta} + \boldsymbol{\xi}_\theta) \approx \boldsymbol{B}(\boldsymbol{\theta})$ is described by*

$$\pi(\boldsymbol{x}) = \nabla\tilde{g}\left(\boldsymbol{B}(\boldsymbol{x})^\top \nabla_x V\right), \qquad \boldsymbol{\xi}_\theta = -\boldsymbol{\Delta}_\nu sign\left(z_{\boldsymbol{\theta}}\right) + \boldsymbol{\mu}_\nu$$
$$\text{with } z_\theta = \left(\frac{\partial \boldsymbol{a}(\boldsymbol{x};\ \boldsymbol{\theta})}{\partial \boldsymbol{\theta}} + \frac{\partial \boldsymbol{B}(\boldsymbol{x};\ \boldsymbol{\theta})}{\partial \boldsymbol{\theta}} \pi(\boldsymbol{x})\right)^\top \nabla_x V,$$

the parameter mean $\boldsymbol{\mu}_\nu = (\boldsymbol{\nu}_{max} + \boldsymbol{\nu}_{min})/2$ and parameter range $\boldsymbol{\Delta}_\nu = (\boldsymbol{\nu}_{max} - \boldsymbol{\nu}_{min})/2$.

Proof Sketch Theorem 2.1 For the admissible set Ω_E, Eq. 4.9 can be written with the explicit constraint. This optimization is described by

$$\begin{aligned}\rho V^* &= \max_{\boldsymbol{u}} \min_{\boldsymbol{\xi}_x} r(\boldsymbol{x}, \boldsymbol{u}) + f(\boldsymbol{x}, \boldsymbol{u}, \boldsymbol{\xi}_x)^\top \nabla_x V^* \text{ s.t. } \boldsymbol{\xi}_x^\top \boldsymbol{\xi}_x \le \alpha^2, \\ &= \max_{\boldsymbol{u}} \min_{\boldsymbol{\xi}_x} q_c(\boldsymbol{x}) - g_c(\boldsymbol{u}) + \left[\boldsymbol{a}(\boldsymbol{x}) + \boldsymbol{B}(\boldsymbol{x})\boldsymbol{u} + \boldsymbol{\xi}_x\right]^\top \nabla_x V^*.\end{aligned}$$

Therefore, the optimal action is described by

$$\boldsymbol{u}_t = \arg\max_{\boldsymbol{u}} \nabla_x V^\top \boldsymbol{B}\, \boldsymbol{u} - g_c(\boldsymbol{u}) \;\Rightarrow\; \boldsymbol{u}_t = \nabla\tilde{g}_c\left(\boldsymbol{B}^\top \nabla_x V\right).$$

The optimal state disturbance is described by

$$\boldsymbol{\xi}_x^* = \arg\min_{\boldsymbol{\xi}_x} \nabla_x V^\top \boldsymbol{\xi}_x \quad \text{s.t.} \quad \frac{1}{2}\left[\boldsymbol{\xi}_x^\top \boldsymbol{\xi}_x - \alpha^2\right] \le 0.$$

This constrained optimization can be solved using the Karush–Kuhn–Tucker (KKT) conditions. The resulting optimal adversarial state perturbation is described by

$$\boldsymbol{\xi}_x = -\alpha \frac{\nabla_x V}{\|\nabla_x V\|_2}.$$

□

Proof Sketch Theorem 2.4 Equation 4.9 can be written as

$$\rho V^* = \max_{\boldsymbol{u}} \min_{\boldsymbol{\xi}_x} r(\boldsymbol{x}, \boldsymbol{u}) + f(.)^\top \nabla_x V^* \text{ s.t. } \left(\boldsymbol{\xi}_\theta - \boldsymbol{\mu}_\nu\right)^2 \le \boldsymbol{\Delta}_\nu^2$$

by replacing the admissible set Ω_A with an explicit constraint. In the following we abbreviate $\boldsymbol{B}(\boldsymbol{x}; \boldsymbol{\theta} + \boldsymbol{\xi}_\theta)$ as $\boldsymbol{B}_\xi$ and $\boldsymbol{a}(\boldsymbol{x}; \boldsymbol{\theta} + \boldsymbol{\xi}_\theta)$ as $\boldsymbol{a}_\xi$. Substituting Eqs. 4.5 and 4.13 simplifies the optimization to

$$\boldsymbol{u}^*, \boldsymbol{\xi}_\theta^* = \arg\max_{\boldsymbol{u}} \arg\min_{\boldsymbol{\xi}} \left[\left(\boldsymbol{a}_\xi + \boldsymbol{B}_\xi \boldsymbol{u}\right)^\top \nabla_x V^* - g_c(\boldsymbol{u}) \right].$$

This nested max-min optimization can be solved by first solving the inner optimization w.r.t. to $\boldsymbol{u}$ and substituting this solution into the outer maximization. The Lagrangian for the optimal model disturbance is described by

$$\boldsymbol{\xi}^* = \arg\min_{\boldsymbol{\xi}} \ \left(\boldsymbol{a}_\xi + \boldsymbol{B}_\xi \boldsymbol{u}\right)^\top \nabla_x V^* + \frac{1}{2} \boldsymbol{\lambda}^\top \left((\boldsymbol{\xi}_\theta - \boldsymbol{\mu}_\nu)^2 - \boldsymbol{\Delta}_\nu^2 \right).$$

Using the KKT conditions this optimization can be solved. The stationarity condition yields

$$z_\theta + \boldsymbol{\lambda}^\top (\boldsymbol{\xi}_\theta - \boldsymbol{\mu}_\nu) = 0 \qquad \Rightarrow \qquad \boldsymbol{\xi}_\theta^* = -z_\theta \oslash \boldsymbol{\lambda} + \boldsymbol{\mu}_\nu$$

with the elementwise division $\oslash$. Using the primal feasibility and the complementary slackness, the optimal $\boldsymbol{\lambda}^*$ can be computed. The resulting optimal model disturbance is described by

$$\boldsymbol{\xi}_\theta^*(\boldsymbol{u}) = -\boldsymbol{\Delta}_\nu sign\big(z_\theta(\boldsymbol{u})\big) + \boldsymbol{\mu}_\nu$$

as $z_\theta \oslash \|z_\theta\|_1 = sign(z_\theta)$. The action can be computed by

$$\boldsymbol{u}^* = \arg\max_{\boldsymbol{u}} \nabla_x V^\top \left[\boldsymbol{a}(\boldsymbol{\xi}_\theta^*(\boldsymbol{u})) + \boldsymbol{B}\left(\boldsymbol{\xi}_\theta^*(\boldsymbol{u})\right) \boldsymbol{u} \right] - g_c(\boldsymbol{u}).$$

Due to the envelope theorem [14], the extrema is described by

$$\boldsymbol{B}(\boldsymbol{x}; \boldsymbol{\theta} + \boldsymbol{\xi}_\theta^*(\boldsymbol{u}))^\top \nabla_x V - g_c(\boldsymbol{u}) = 0.$$

This expression cannot be solved without approximation as $\boldsymbol{B}$ does not necessarily be invertible w.r.t. $\boldsymbol{\theta}$. Approximating $\boldsymbol{B}(\boldsymbol{x}; \boldsymbol{\theta} + \boldsymbol{\xi}^*(\boldsymbol{u})) \approx \boldsymbol{B}(\boldsymbol{x}; \boldsymbol{\theta})$, lets one solve for $\boldsymbol{u}$. In this case the optimal action $\boldsymbol{u}^*$ is described by $\boldsymbol{u}^* = \nabla \tilde{g}(\boldsymbol{B}(\boldsymbol{x}; \boldsymbol{\theta})^\top \nabla_x V)$. This common approximation [4] implies that neither agent or the adversary can react to the action of the other and must choose simultaneously. □

4.4 Continuous Fitted Value Iteration

The previous sections showed that the optimizations contained within the HJB and HJI can be solved in closed form. Leveraging these insights, we will derive the proposed algorithms to solve the HJB and HJI using fitted value iteration.

Substituting the optimal actions and adversary, simplifies the Eqs. 4.4 and 4.9 to a differential equations without optimization. These equations are described by

$$\rho\, V^*(\boldsymbol{x}) = r(\boldsymbol{x}, \boldsymbol{u}^*) \;\; + f_c(\boldsymbol{x}, \boldsymbol{u}^*)^\top\, \nabla_{\boldsymbol{x}} V^*,$$
$$\rho\, V^*(\boldsymbol{x}) = r(\boldsymbol{x}, \boldsymbol{u}^*) \;\; + f_c(\boldsymbol{x}, \boldsymbol{u}^*, \boldsymbol{\xi}^*)^\top\, \nabla_{\boldsymbol{x}} V^*,$$

with the optimal action $\boldsymbol{u}^*$ and optimal adversary $\boldsymbol{\xi}^*$ described by Eq. 4.16. Within the machine learning community various approaches have been proposed to solve these differential equations using standard regression techniques [21, 57, 79], including our previous work [49]. The problem with these approaches is, that these differential equation do not have a unique solutions without considering the boundary constraint, which implies that the optimal action always prevent the system to leave the state domain. The boundary condition is described by

$$f(\bar{\boldsymbol{x}}, \boldsymbol{u}^*)^\top \eta(\bar{\boldsymbol{x}}) \leq 0 \qquad \text{for } \bar{\boldsymbol{x}} \in \partial\mathcal{X} \tag{4.19}$$

with the outward pointing normal vector η defined on the state domain boundary $\partial\mathcal{X}$ makes the solution unique [26]. In the case of LQR, this boundary condition implies the positive-definiteness of the quadratic value function. Incorporating this boundary constraint within the optimization problem is challenging, as the state domain boundary is unknown and the commonly used black-box function approximators are local, hence incoporating the boundary constraint within the optimization does not ensure a globally coherent solution. To overcome this shortcoming the existing regression approaches use additional optimization tricks and specific value function representations.

To overcome the problems arising from the boundary condition when solving the HJB using regression techniques, we use fitted value iteration (FVI) [3, 10, 58, 84]. FVI is an extension of the classical dynamic programming Value iteration (VI) [6] to continuous states using a function approximator. Value iteration is preferable as value iteration based techniques guarantee to converge to the desired optimal value function when ignoring the approximation error arising from function approximation [58]. Therefore, it is not necessary to incorporate the boundary constraint required for regression techniques. FVI iteratively computes the value function target and minimizes the ℓ_p-norm between the target and the approximation $V^k(\boldsymbol{x};\ \psi)$ until the value function has converged. Mathematically, this approach is described by

$$V_{\text{tar}}(\boldsymbol{x}_t) = \max_{\boldsymbol{u}} r(\boldsymbol{x}_t, \boldsymbol{u}) + \gamma V^k(\boldsymbol{x}_{t+1};\ \psi_k), \tag{4.20}$$

$$\psi_{k+1} = \arg\min_{\psi} \sum_{\boldsymbol{x}\in\mathcal{D}} \| V_{\text{tar}}(\boldsymbol{x}) - V^k(\boldsymbol{x};\ \psi)\|_p^p \tag{4.21}$$

with the parameters ψ_k at iteration k and the fixed dataset $\mathcal{D}$. While for discrete states and actions for $\gamma < 1$, VI is proven to converge to the optimal value function [68], this convergence proof of VI does not generalize to FVI as the fitting of the value target is not necessarily a contraction [3, 10, 58, 84]. However, empirically this approach has been successfully used to retrieve the optimal value function or Q function [3, 10, 22, 23, 56, 58, 73, 80, 84]. To solve the HJB and HJI, the value function target must be adapted and is described by

$$V_{\text{tar}}(\boldsymbol{x}_t) = r\left(\boldsymbol{x}_t, \boldsymbol{u}^*\right) + \gamma V^k(\boldsymbol{x}_{t+1};\ \psi_k),$$
$$\text{with} \quad \boldsymbol{x}_{t+1} = \boldsymbol{x}_t + \int_t^{t+\Delta T} f_c\left(\boldsymbol{x}_\tau, \boldsymbol{u}_\tau^*, \boldsymbol{\xi}_\tau^*\right) d\tau,$$

and the time step ΔT. The selection of the time step is important as this discretization affects the quality of the value approximation and the convergence speed. The convergence speed decreases exponentially with smaller time step, as the contraction coefficient of VI is proportional to γ and γ approaches 1 with a smaller time step, i.e., $\gamma = \lim_{\Delta t \to 0} \exp(-\rho \Delta t) = 1$. This slower convergence is intuitive as the smaller time step increases the number of steps through which the information of high value function states must be propagated using the recursive value function update. Therefore, one has to balance a small time step to improve the approximation accuracy and a large time step to improve convergence speed. In practice, this trade-off is solved by starting with a large time step and decreasing the time step until the approximation of the value function is sufficient. Finding an optimal time step for all systems is not possible as the time step depends on the frequency response on the specific dynamical system.

4.4.1 N-Step Value Function Target

To further improve the convergence speed of fitted value iteration, the exponentially weighted n-step value function target is used. This value target is described by

$$V_{\text{tar}}(\boldsymbol{x}) = \int_0^T \beta \exp(-\beta t)\, R_t\, dt + \exp(-\beta T) R_T,$$

$$R_t = \int_0^t \exp(-\rho\tau)\, r_c(\boldsymbol{x}_\tau, \boldsymbol{u}_\tau) d\tau + \exp(-\rho t) V^k(\boldsymbol{x}_t),$$

and the exponential decay constant β, can be used. This approach is the continuous-time counterpart of the discrete eligibility trace of TD(λ) with $\lambda = \exp(-\beta \Delta t)$ [76]. With respect to deep RL, this discounted n-step value target is similar to the generalized advantage estimation (GAE) of PPO [74, 75] and model-based value expansion (MVE) [12, 24]. GAE and MVE have shown that the n-step target increases the sample efficiency and lead to faster convergence to the optimal policy. The integrals can be solved using any ordinary differential equation solver with fixed or adaptive step-size. We use the explicit Euler integrator with fixed steps to solve the integral for all samples in parallel using batched operations on the GPU. The nested integrals can be computed efficiently by recursively splitting the integral and reusing the estimate of the previous step. In practice we treat β as a hyperparameter and select T such that the weight of the R_T is $\exp(-\beta T) = 10^{-4}$.

4.4.2 Dataset

Equation 4.21 fits the value function using the dataset $\mathcal{D}$. This dataset can be fixed as in dynamic programming or offline/batch RL or a replay memory containing the visited states of the current policy π^k. We refer to the latter as real-time dynamic programming (RTDP) as Barto et al. [5] introduced the online version of dynamic programming first. In the case of the fixed dataset, the dataset can either originate from a previous learning process, which is frequently used in the offline RL benchmarks [30], or uniformly sampled from the state domain $\mathcal{X}$. Within this work, we sample uniformly from the state domain as the used state dimensionality is low-dimensional. For RTDP, the dataset is a replay memory containing the visited states of the current policy as in most modern deep reinforcement learning algorithms. In this case, the exploration of the policy is important as the policy needs to cover the state space to discover high reward configurations. In the offline case, no exploration is needed.

4.4.3 Admissible Set

For the state, action, and observation adversary the signal energy is bounded. We limit the energy of $\boldsymbol{\xi}_x$, $\boldsymbol{\xi}_u$ and $\boldsymbol{\xi}_o$ as the non-adversarial disturbances are commonly modeled as multivariate Gaussian distribution. Therefore, the average energy is determined by the noise covariance matrix. For the model parameters $\boldsymbol{\theta}$ a common practice is to assume that the approximate model parameters have a model error of up to $\pm 15\%$ [59, 60]. Hence, we bound the amplitude of each component. To not overfit to the deterministic worst-case system of V and enable the discovery of good actions, the amplitude of the adversarial actions of ξ_x, ξ_u, ξ_o is modulated using a Wiener process. This random process allows a continuous-time formulation that is agnostic to the sampling frequency.

Algorithm 4.1 Robust Fitted Value Iteration (rFVI)

Input: Model $f_c(\boldsymbol{x}, \boldsymbol{u})$, Dataset $\mathcal{D}$ & Admissible Set Ω_ξ
Result: Value Function $V^*(\boldsymbol{x};\ \psi^*)$
while not converged **do**
  // Compute Value Target for $\boldsymbol{x} \in \mathcal{D}$:
  $\boldsymbol{x}_\tau = \boldsymbol{x}_i + \int_0^\tau f_c(\boldsymbol{x}_t, \boldsymbol{u}_t, \xi_t^x, \xi_t^u, \xi_t^o, \xi_t^\theta)dt$
  $R_t = \int_0^t \exp(-\rho\tau)\, r_c(\boldsymbol{x}_\tau, \boldsymbol{u}_\tau)d\tau + \exp(-\rho t)V^k(\boldsymbol{x}_t)$
  $V_{\text{tar}}(\boldsymbol{x}_i) = \int_0^T \beta\ \exp(-\beta t)\ R_t\ dt + \exp(-\beta T)R_T$

  // Fit Value Function:
  $\psi_{k+1} = \arg\min_\psi \sum_{\boldsymbol{x}\in\mathcal{D}} \| V_{\text{tar}}(\boldsymbol{x}) - V(\boldsymbol{x};\ \psi)\|^p$

  if RTDP rFVI **then**
    // Add samples from π^{k+1} to FIFO buffer $\mathcal{D}$
    $\mathcal{D}^{k+1} = h(\mathcal{D}^k, \{\boldsymbol{x}_0^{k+1}\ \dots\ \boldsymbol{x}_N^{k+1}\})$
  end if
end while

Combining value iteration with the analytic optimal policy and adversary yields the two algorithms cFVI [51] and rFVI [50]. While cFVI is used to solve the HJB, rFVI is used to solve the HJI. We refer to these algorithms as an extension of value iterations as these algorithms also extend FVI approach to continuous actions and adversarial RL, which was previously not possible. Previously FVI was limited to discrete actions. Furthermore, we differentiate between a dynamic programming version of the algorithm using a fixed dataset, e.g., DP cFVI and DP rFVI and the online version using a replay memory, i.e., RTDP cFVI and RTDP rFVI. The algorithms are summarized in Algorithm 4.1.

4.4.4 Value Function Representation

For the value function representation one can use any differentiable black-box function approximator. One common choice is a feed-forward network, i.e., Multi-Layer Perceptron (MLP), as this approximator enables an efficient computation of the value function gradient w.r.t. the network inputs.

While the standard network architectures are sufficient, one can improve the performance by leveraging insights from the common control cost choices to structure the architecture. These structured representations are preferable as these limit the hypothesis space of the representable value functions. For continuous control tasks, the state reward is often a negative distance measure between $\boldsymbol{x}_t$ and the desired state $\boldsymbol{x}_{\text{des}}$. Hence, q_c is negative definite, i.e., $q(\boldsymbol{x}) < 0\ \forall\ \boldsymbol{x} \neq \boldsymbol{x}_{\text{des}}$ and $q(\boldsymbol{x}_{\text{des}}) = 0$. These properties imply that V^* is a negative Lyapunov function, as V^* is negative definite, $V^*(\boldsymbol{x}_{\text{des}}) = 0$ and $\nabla_x V^*(\boldsymbol{x}_{\text{des}}) = \boldsymbol{0}$ [39]. With a deep network a similar representation can be achieved by

$$V(\boldsymbol{x};\ \psi) = -\left(\boldsymbol{x} - \boldsymbol{x}_{\text{des}}\right)^{\top} \boldsymbol{L}(\boldsymbol{x};\ \psi)\boldsymbol{L}(\boldsymbol{x};\ \psi)^{\top} \left(\boldsymbol{x} - \boldsymbol{x}_{\text{des}}\right)$$

with $\boldsymbol{L}$ being a lower triangular matrix with positive diagonal. This positive diagonal ensures that $\boldsymbol{L}\boldsymbol{L}^{\top}$ is positive definite. Simply applying a ReLu activation to the predicted value function instead of using the local quadratic representation to achieve the negative definiteness is not sufficient. This ReLu transformation cannot guarantee $\nabla_x V^*(\boldsymbol{x}_{\text{des}}) = \boldsymbol{0}$ and would zero the gradient for all positive values, which make the optimization of the value function harder. The used local quadratic representation guarantees that the gradient and hence, the action, is zero at the desired state. However, this representation can also not guarantee that the value function has only a single extrema at $\boldsymbol{x}_{\text{des}}$ as required by the Lyapunov theory. In practice, the local regularization of the quadratic structure to avoid high curvature approximations is sufficient as the global structure is defined by the value function target. $\boldsymbol{L}$ is the mean of a deep network ensemble with N independent parameters ψ_i. The ensemble mean smoothes the initial value function and is differentiable. Similar representations have been used by prior works in the safe reinforcement learning community [7, 8, 15, 29, 43, 72]. It is important to point out that this network architecture is different from NAF [29] as NAF uses a Q-function that is quadratic w.r.t. the actions while we use a value function that is quadratic w.r.t. to the state.

Additional state transformations can be incorporated into the value function to enable easier representations. For example, the standard feature transform for a continuous revolute joint maps the joint state $\boldsymbol{x} = [\theta,\ \dot{\theta}]$ to $z = [\sin(\theta),\ \cos(\theta),\ \dot{\theta}]$ can be incorporated to avoid the discontinuity at $\pm\pi$. In this case the transformed state z lies on the tube shaped manifold. Therefore, the value function gradient must be projected into the tangent space, which is not guaranteed when using deep networks. For the state transformation $h(\boldsymbol{x})$ with $V(x;\ \psi) = f(h(\boldsymbol{x});\ \psi)$ this projection is described by $\nabla_x V(x;\ \psi) = \partial f(h(\boldsymbol{x});\ \psi)/\partial h\ \ \partial h(\boldsymbol{x})/\partial \boldsymbol{x}$ and the gradient points in a sensible direction.

4.5 Experiments

In the non-linear control experiments, we apply cFVI and rFVI to control under-actuated systems. The sim2real experiments test the policy robustness by transferring the learned policy to the physical system and compare their performance to the standard deep RL approaches. More precisely, we want to answer the following questions:

Q1: Can cFVI & rFVI obtain the optimal policies that control the simulated system?

Q2: What are the qualitative differences between the policies obtained by cFVI & rFVI?

Q3: Does the n-step value target improved the convergence speed of the optimal policy?

Q4: How does the admissible set of adversaries affect the performance of the optimal policy?

Q5: Is the locally quadratic value function architecture beneficial compared to a standard feed-forward network?

Q6: Are the obtained policies robust enough to be transferred to the real system with varying physical parameters?

4.5.1 Experimental Setup

To answer these research questions, we apply the proposed algorithms to non-linear sim2real control of under-actuated systems and compare the performance to standard actor-critic deep RL approaches. The detailed experiment description including the hyperparameters is contained in the appendix. The code of cFVI and rFVI is available at https://github.com/milutter/value_iteration.

Systems. The performance of the algorithms is evaluated using the *swing-up* the torque-limited pendulum, cartpole and Furuta pendulum. The physical cartpole and Furuta pendulum are manufactured by Quanser [69]. For simulation, we use the equations of motion and physical parameters of the supplier. Both systems have very different characteristics. The Furuta pendulum consists of a small and light pendulum (24 g, 12.9 cm) with a strong direct-drive motor. Even minor differences in the action cause large changes in acceleration due to the large amplification of the mass-matrix inverse. Therefore, the main source of uncertainty for this system is the uncertainty of the model parameters. The cartpole has a longer and heavier pendulum (127 g, 33.6 cm). The cart is actuated by a geared cogwheel drive. Due to the larger masses the cartpole is not so sensitive to the model parameters. The main source of uncertainty for this system is the friction and the backlash of the linear actuator. The systems are simulated and observed with 500 Hz. The control frequency varies between algorithm and is treated as hyperparameter.

To test the robustness to changes in the dynamics parameters, we glue additional weights to the pendulum. If we add negative weights to the pendulum, we attach the weights to the opposite lever of the pendulum. This moves the center of mass of the pendulum closer to the rotary axis. Therefore, this shift reduces the downward force and is equivalent to a lower pendulum mass.

Baselines. The performance is compared to the actor-critic deep RL methods: DDPG [46], SAC [31] and PPO [75]. The robustness evaluation is only performed for the best performing baselines on the nominal physical system. The initial state distribution is abbreviated by {SAC, PPO, DDPG}-U for a uniform distribution of the pendulum angle and {SAC, PPO, DDPG}-N for a Gaussian distribution. The baselines with Gaussian initial state distribution did not achieve robust performance on the nominal system. If the baseline uses uniform domain randomization the acronym is appended with UDR. To enable a fair comparison between these different approaches, we select the time step for each baseline is determined using a hyperparameter sweep and only compare the converged policies. Selecting different time steps for each baseline is important as the many deep RL algorithms struggle with small time steps [77]. We only compare the final policies as the optimization procedure between the actor-critic approaches and the proposed techniques are different. Therefore, the learning speed between these methods cannot be compared.

Reward Function. The desired state for all tasks is the upward pointing pendulum at $\boldsymbol{x}_{\text{des}} = \mathbf{0}$. The state reward is described by $q_c(\boldsymbol{x}) = -(\boldsymbol{z} - \boldsymbol{z}_{\text{des}})^T \boldsymbol{Q}(\boldsymbol{z} - \boldsymbol{z}_{\text{des}})$ with the positive definite matrix $\boldsymbol{Q}$ and the transformed state $\boldsymbol{z}$. For continuous joints the joint state is transformed to $z_i = \pi^2 \sin(x_i)$. The action cost is described by $g_c(\boldsymbol{u}) = -2\, \boldsymbol{\beta}\, \boldsymbol{u}_{\max}/\pi \log\cos(\pi\, \boldsymbol{u}/(2\, \boldsymbol{u}_{\max}))$ with the actuation limit $\boldsymbol{u}_{\max}$ and the positive constant $\boldsymbol{\beta}$. This barrier shaped cost bounds the optimal actions. The corresponding policy is shaped by $\nabla\tilde{g}(\boldsymbol{w}) = 2\, \boldsymbol{u}_{\max}/\pi\, \tan^{-1}(\boldsymbol{w}/\boldsymbol{\beta})$. For the experiments, the reward parameters are

$$\begin{aligned} \text{Pendulum:} \quad & \boldsymbol{Q}_{\text{diag}} = [01.0,\ 0.1]\,, & \beta = 0.5 \\ \text{Cartpole:} \quad & \boldsymbol{Q}_{\text{diag}} = [25.0,\ 1.0,\ 0.5,\ 0.1]\,, & \beta = 0.1 \\ \text{Furuta Pendulum:} \quad & \boldsymbol{Q}_{\text{diag}} = [01.0,\ 5.0,\ 0.1,\ 0.1]\,, & \beta = 0.1 \end{aligned}$$

Evaluation. To evaluate rFVI and the baselines we separately compare the state and action reward as these algorithms optimize a different objective. Hence, these algorithms trade-off state and action associated rewards differently. It is expected that the worst-case optimization uses higher actions to prevent deviation from the optimal trajectory. On the physical system, the performance is evaluated using the 25th, 50th, and 75th reward percentile as the reward distribution is multi-modal. The swing-up tasks are considered solved if the pendulum angle α is below $\pm 5°$ degree for the last second. The reported average return is the accumulated reward for one trajectory averaged over 100 roll outs in simulation and 15 roll outs on the real system. The learning curves for the four proposed algorithms are averaged over 20 seeds and show the min/max variance between seeds (Fig. 4.2). For the different ablation studies (Figs. 4.3, 4.4 and 4.5), the learning curves are averaged over 5 seeds and show the min/max variance between seeds.

4.5.2 *Experimental Results*

For each of the research questions we summarize the empirical results and answer the question in the respective section. Videos of all performed experiments are available at https://sites.google.com/view/rfvi.

Q1 Control Performance. The learning curves for the three dynamical systems are shown in Fig. 4.2. The quantitative comparison to the baselines is summarized in Table 4.3. DP cFVI, DP rFVI and RTDP cFVI obtain a policy that performs the swing-up and balances the pendulum. Only RTDP rFVI does not obtain a successful policy

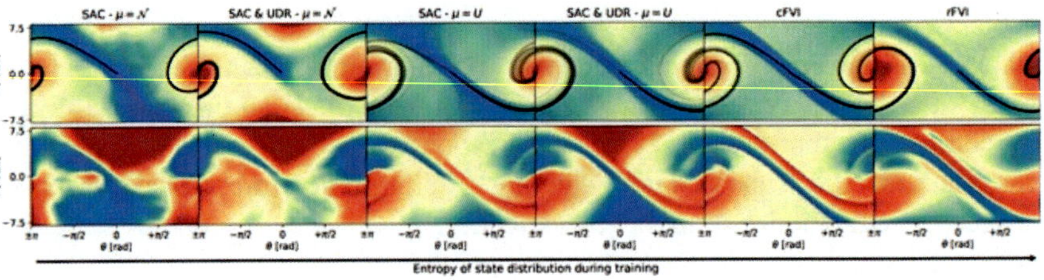

Fig. 4.1 The optimal Value function V^* and policy π^* of rFVI, cFVI, and four different variations of SAC. All policies achieve nearly identical reward on the nominal dynamics model. The variations of SAC demonstrate the change of the policy when increasing the entropy of the state distribution during training. The entropy is increased by enlarging the initial state distribution μ and using domain randomization. For SAC and $\mu = \mathcal{N}(\pm\pi, \sigma)$ the optimal policy is only valid on the optimal trajectory. For SAC UDR and $\mu = \mathcal{U}(-\pi, +pi)$, the policy is applicable on the complete state domain. rFVI and cFVI perform value iteration on the compact state domain and naturally obtain an optimal policy applicable on the complete state-domain. rFVI adapts V^* and π^* to have a smaller ridge leading up to the upright pendulum and exerts higher actions when deviating from the optimal trajectory

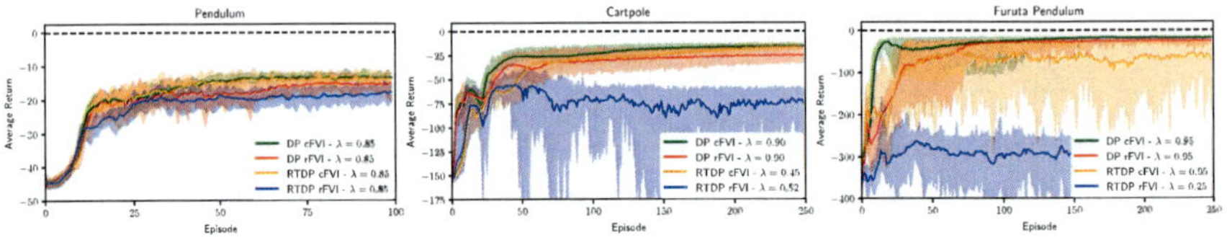

Fig. 4.2 The learning curves for DP rFVI, DP cFVI, RTDP cFVI, and RTDP rFVI averaged over 20 seeds. The shaded area displays the *min/max* range between seeds. DP rFVI learns slower compared to DP cFVI on the carpole and Furuta pendulum as the adversary prevents learning. RTDP rFVI does not learn the task as the adversary is too strong for the online variant of rFVI despite using the identical admissible set as the offline variant DP rFVI

for the cartpole and the Furuta pendulum (See Q4 for additional details). DP cFVI learns the fastest compared to the other variants. RTDP cFVI learns slower due to the required exploration while for DP rFVI the adversary slows down the convergence to the optimal value function. Quantitatively, the DP cFVI performs comparably to the best performing deep RL algorithms. DP rFVI obtains a lower reward compared to DP cFVI due to the adversary. This reward difference is expected rFVI minimizes the risk and hence, selects a more conservative solution with lower reward.

Q2 Policy Difference. The main difference between the policies obtained by cFVI and rFVI is that the robust variant converges to a stiffer policy. This stiffer policy exerts higher actions as soon as the system state leaves the optimal trajectory. This behavior is caused due to the adversary, which frequently perturbs the system state to leave the optimal trajectory. This difference can be visualized for the pendulum (Fig. 4.1). For the cFVI policy the color gradient is much smoother, while for the rFVI policy the color gradient changes abrupt between the maximum actions. Therefore, the rFVI policy performs close to bang-bang control. Furthermore, the ridge leading

Table 4.3 Average return on the simulated and physical systems. The average ranking describes the decrease in reward compared to the best result averaged on all systems. Therefore, a small decrease shows that the algorithm performs close to the best algorithm on each system. The initial state distribution during training is noted by μ. The dynamics are either deterministic model $\theta \sim \delta(\theta)$ or sampled using uniform domain randomization $\theta \sim \mathcal{U}(\theta)$. During evaluation the roll outs start with the pendulum pointing downwards

Algorithm	μ	θ	Simulated pendulum		Simulated cartpole		Simulated furuta pendulum	
			Success (%)	Return ($\mu \pm 2\sigma$)	Success (%)	Return ($\mu \pm 2\sigma$)	Success (%)	Return
DP rFVI (ours)	–	$\delta(\theta)$	100.0	−032.7 ± 000.3	100.0	−027.1 ± 004.8	100.0	−041.3 ± 010.8
DP cFVI (ours)	–	$\delta(\theta)$	100.0	**−030.5 ± 000.8**	100.0	**−024.2 ± 002.1**	100.0	**−027.7 ± 001.6**
RTDP cFVI (ours)	$\mathcal{U}$	$\delta(\theta)$	100.0	**−031.1 ± 001.4**	100.0	**−024.9 ± 001.6**	100.0	−040.1 ± 002.7
SAC	$\mathcal{N}$	$\mathcal{U}(\theta)$	100.0	**−031.1 ± 000.1**	100.0	−026.9 ± 003.2	100.0	−029.3 ± 001.5
SAC & UDR	$\mathcal{N}$	$\delta(\theta))$	100.0	−032.9 ± 000.6	100.0	−029.7 ± 004.6	100.0	−032.0 ± 001.1
SAC	$\mathcal{U}$	$\mathcal{U}(\theta)$	100.0	**−030.6 ± 001.4**	100.0	**−024.2 ± 001.4**	100.0	**−028.1 ± 002.0**
SAC & UDR	$\mathcal{U}$	$\mathcal{U}(\theta)$	100.0	−031.4 ± 002.5	100.0	**−024.2 ± 001.3**	100.0	**−028.1 ± 001.3**
DDPG	$\mathcal{N}$	$\mathcal{U}(\theta)$	100.0	**−031.1 ± 000.4**	98.0	−050.4 ± 285.6	100.0	−030.5 ± 003.5
DDPG & UDR	$\mathcal{N}$	$\delta(\theta))$	100.0	−032.5 ± 000.5	100.0	−027.4 ± 002.3	100.0	−034.6 ± 009.8
DDPG	$\mathcal{U}$	$\mathcal{U}(\theta)$	100.0	−031.5 ± 000.7	100.0	−028.2 ± 005.5	100.0	−030.0 ± 001.7
DDPG & UDR	$\mathcal{U}$	$\mathcal{U}(\theta)$	100.0	−032.5 ± 003.6	100.0	−027.2 ± 001.0	100.0	−032.1 ± 001.5
PPO	$\mathcal{N}$	$\mathcal{U}(\theta)$	100.0	−032.0 ± 000.2	100.0	−031.5 ± 007.2	100.0	−081.1 ± 018.3
PPO & UDR	$\mathcal{N}$	$\delta(\theta))$	100.0	−032.3 ± 000.6	100.0	−084.0 ± 007.8	100.0	−040.9 ± 004.6
PPO	$\mathcal{U}$	$\mathcal{U}(\theta)$	100.0	−033.4 ± 004.7	99.0	−039.7 ± 045.7	100.0	−038.2 ± 013.1
PPO & UDR	$\mathcal{U}$	$\mathcal{U}(\theta)$	100.0	−035.6 ± 003.1	100.0	−044.8 ± 021.4	100.0	−048.5 ± 006.2

(continued)

Table 4.3 (continued)

Algorithm	μ	θ	Physical cartpole		Physical furuta pendulum		Average
			Success (%)	Return ($\mu \pm 2\sigma$)	Success (%)	Return ($\mu \pm 2\sigma$)	Return (μ)
DP rFVI (ours)	–	$\delta(\theta)$	100.0	**−074.1 ± 040.3**	100.0	−278.0 ± 034.3	−062.7
DP cFVI (ours)	–	$\delta(\theta)$	73.3	−143.7 ± 210.4	100.0	**−082.1 ± 007.6**	**−019.2**
RTDP cFVI (ours)	$\mathcal{U}$	$\delta(\theta)$	100.0	−101.1 ± 029.0	00.0	−1009.9 ± 004.5	−247.7
SAC	$\mathcal{N}$	$\mathcal{U}(\theta)$	00.0	−518.6 ± 028.1	86.7	−330.7 ± 799.0	−185.8
SAC & UDR	$\mathcal{N}$	$\delta(\theta))$	100.0	−394.8 ± 382.8	100.0	−181.4 ± 157.9	−120.8
SAC	$\mathcal{U}$	$\mathcal{U}(\theta)$	53.3	−144.5 ± 204.0	100.0	−350.8 ± 433.3	−086.5
SAC & UDR	$\mathcal{U}$	$\mathcal{U}(\theta)$	40.0	−296.4 ± 418.9	100.0	−092.3 ± 064.1	−063.8
DDPG	$\mathcal{N}$	$\mathcal{U}(\theta)$	06.7	−536.7 ± 262.7	46.7	−614.1 ± 597.8	−281.4
DDPG & UDR	$\mathcal{N}$	$\delta(\theta))$	00.0	−517.9 ± 117.6	86.7	−192.7 ± 404.8	−156.6
DDPG	$\mathcal{U}$	$\mathcal{U}(\theta)$	06.7	−459.4 ± 248.3	100.0	−146.6 ± 218.3	−126.0
DDPG & UDR	$\mathcal{U}$	$\mathcal{U}(\theta)$	00.0	−318.1 ± 063.4	100.0	−156.7 ± 246.4	−091.7
PPO	$\mathcal{N}$	$\mathcal{U}(\theta)$	00.0	−287.9 ± 068.8	33.3	−718.7 ± 456.1	−261.7
PPO & UDR	$\mathcal{N}$	$\delta(\theta))$	00.0	−435.4 ± 111.9	46.7	−935.7 ± 711.6	−370.0
PPO	$\mathcal{U}$	$\mathcal{U}(\theta)$	00.0	−183.8 ± 018.0	60.0	−755.3 ± 811.0	−219.4
PPO & UDR	$\mathcal{U}$	$\mathcal{U}(\theta)$	40.0	−143.8 ± 016.1	100.0	**−080.6 ± 010.8**	−054.4

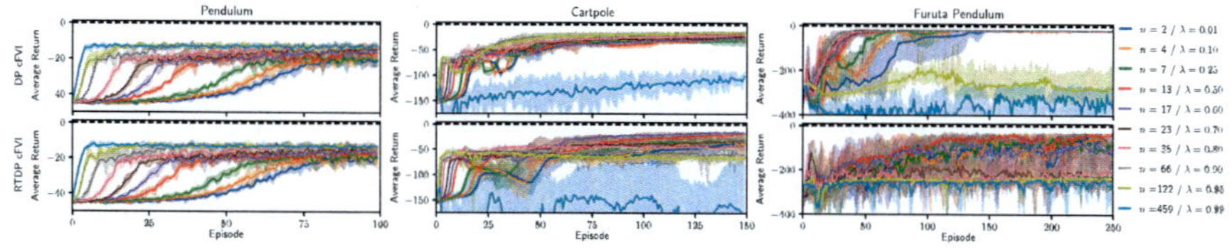

Fig. 4.3 The learning curves averaged over 5 seeds for the n-step value function target. The shaded area displays the *min/max* range between seeds. The step count is selected such that $\lambda^n = 10^{-4}$. Increasing the horizon of the value function target increases the convergence rate to the optimal value function. For very long horizons the learning diverges as it over fits to the current value function approximation. Furthermore, the performance of the optimal policy also increases with roll out length

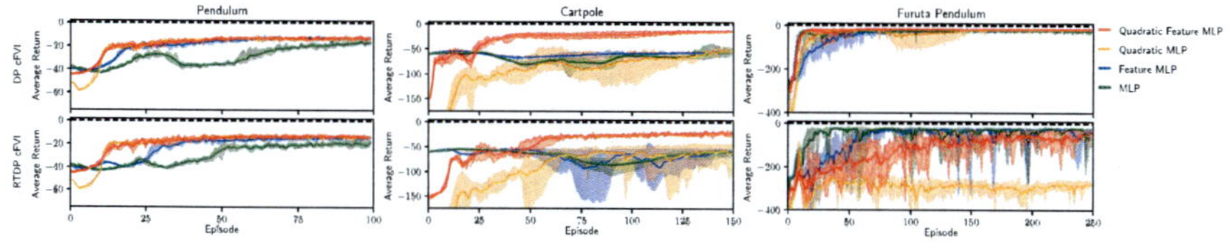

Fig. 4.4 The learning curves averaged over 5 seeds for the different model architectures. The shaded area displays the *min/max* range between seeds. All network architectures are capable of learning the value function and policy for most of the tasks. The locally quadratic network architecture increases learning speed compared to the baselines. The structured architecture acts as an inductive bias that shapes the exploration. The global maximum of the locally quadratic value function is guaranteed at $\boldsymbol{x}_{\text{des}}$ and hence the initial policy performs hill-climbing towards this point

up to the balancing point in the center is much smaller for rFVI as the policy expects the adversary to push the state off the cliff, leading to a much lower reward. Therefore, the rFVI policy is more conservative and uses a larger safety margin.

Q3 N-Step Value Target. The learning curves for varying n-steps is shown in Fig. 4.3. The ablation study shows that the convergence speed increases when increasing the number of steps. It is important to point out that the learning speed increases w.r.t. to the number of episodes. In terms of computational cost, the number of steps increases the computational complexity by n as the simulation is sequential. A surprising observation is that the cartpole and Furuta pendulum do not converge for very long trajectories despite using the true model. This degraded performance is due to overfitting to the value function approximation during training. As the value function is randomly initialized at the beginning, using long trajectories leading to potentially untrained regions in the state domain might lead to bad local optima. For the RTDP variant, this effect is amplified as the value function approximation is limited to the current state distribution and long trajectories are prone to leaving the state distribution. Therefore, the optimal step count is lower for RTDP than for DP. Furthermore, we also observed that rFVI performs better when using slightly lower horizons compared to cFVI.

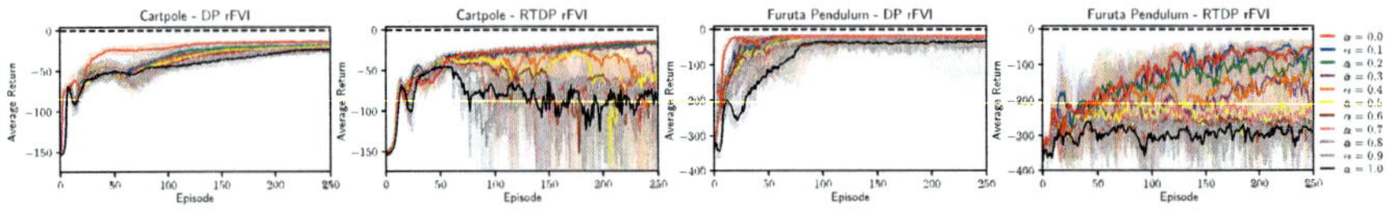

Fig. 4.5 The learning curves for DP rFVI and RTDP rFVI with different adversary amplitudes averaged over 5 seeds. The shaded area displays the *min/max* range between seeds. The α corresponds to the percentage of the admissible set for all adversaries, i.e., with increasing α the adversary becomes more powerful. For DP rFVI the stronger adversaries do affect the final performance only marginally. For RTDP rFVI the adversaries become too powerful for small α and prevent learning of the optimal policy. This effect is especially distinct for the Furuta pendulum as this system is very sensitive due to the low masses. Therefore, DP rFVI can learn a good optimal policy despite very strong adversaries

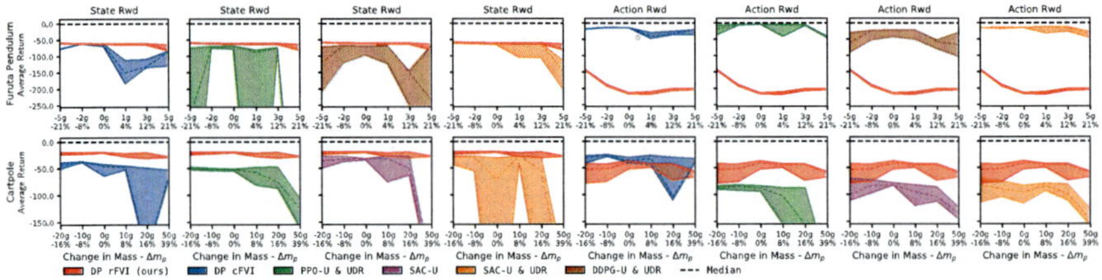

Fig. 4.6 The 25th, 50th, and 75th reward percentile for the physical Furuta pendulum and cartpole with varied pendulum weights. DP rFVI achieves a higher state reward for real-world systems compared to the baselines. For the different weights, the reward remains nearly constant. For the Furuta pendulum the action cost is significantly higher compared to the baselines as the DP rFVI causes a chattering during balancing due to the high actions and minor time delays in the control loop. If only the swing-up phase is considered the rewards are comparable

Q4 Admissible Set. The learning curves for the linearly scaled admissible set are shown in Fig. 4.5. One can observe that making the adversary more powerful, i.e., increasing α, slightly decreases the obtained reward as the policy obtained by DP rFVI is more conservative. Furthermore, the learning speed is decreased. However, for all α the policy learns to complete the task. For RTDP rFVI increasing the admissible set of the adversary has a more significant impact. If the adversary becomes too powerful, the policy does not achieve the task. For example, for the cartpole the reward initially increases but drops for large admissible sets. For the Furuta pendulum the reward does not even increase at the beginning for larger α. This behavior is due to the limited exploration of the policy. Due to the adversary, the policy does not discover that an action sequence exists to achieve the task. Therefore, the policy often converges to a pessimistic solution that does not exert any action. This policy is locally optimal as the adversary will always prevent the policy from completing the task. Hence, performing actions and incurring action penalties is not desirable. This effect is more pronounced for the Furuta pendulum as the system is more sensitive due to its small lengths and masses.

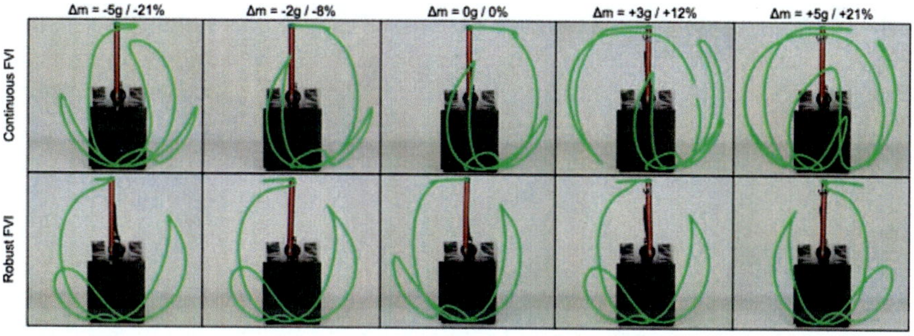

Fig. 4.7 The tracked trajectories for DP rFVI and DP cFVI on the Furuta pendulum for different pendulum weights. The trajectories of rFVI do not significantly change when the pendulum mass is altered. For DP cFVI the trajectories start to change when additional weight is added. For these system dynamics, DP cFVI requires some failed swing-ups until the policy can balance the pendulum

Q5 Network Architecture. The learning curves for different network architectures are shown in Fig. 4.4. The learning curves show that the locally quadratic network architecture using the sine/cosine transform for continuous revolute joints performs the most reliable. While the standard MLP with feature transform performs well for the pendulum and Furuta pendulum, it does not perform well for the cartpole.

Q6 Sim2Real Transfer. The results of the sim2real transfer are summarized in Fig. 4.6 and Table 4.3. And extensive video documentation showing the performance on the physical systems is provided at https://sites.google.com/view/rfvi. Both cFVI and rFVI can be transferred to the physical system and achieve a successful swing-up. On the nominal Furuta pendulum cFVI obtains a higher reward than rFVI, as rFVI uses higher actions. On the nominal cartpole, rFVI has a higher reward and higher success rate compared to cFVI. In terms of robustness w.r.t. changes of the physical parameters, rFVI achieves a reliable swing-up even when weights are added to the pendulum. Therefore, the obtained state reward is not affected by the varied mass (Fig. 4.6). This difference can be nicely observed in Fig. 4.7. The pendulum trajectory of the rFVI policy is identical for all weight configurations. In contrast, the cFVI policy needs multiple unsuccessful tries until the pendulum is upright and balanced for added weights ≥ 3 g. However, during balancing the Furuta pendulum, rFVI performs bang-bang control, which leads to chattering due to minor delays in the control loop. In contrast to rFVI, cFVI keeps the pendulum still when balancing as the policy does not apply so strong actions. For the cartpole, rFVI obtains a robust policies that performs a more consistent swing-up and balancing (Fig. 4.6). The difference between the policies is especially visible during the balancing of the cartpole. Due to the stiff rFVI policy, the large actions immediately break the stiction of the linear actuator. Therefore, the cart is balanced at the center. For the cFVI policy the cart oscillates around the center as the pendulum needs to fall a bit until the deviation is large enough to exert actions that break the stiction.

When comparing the performance of cFVI and rFVI to the deep RL algorithms with uniform domain randomization, the proposed algorithms perform comparable or better. For example on the nominal system, cFVI performs as good as the best deep RL baseline. For example, on the Furuta pendulum most baselines complete the task but cFVI obtains the highest reward and only PPO with domain randomization obtains a similar reward. On the robustness experiments, rFVI performs better than the deep RL baselines with domain randomization. The baselines including domain randomization start to fail when additional weights are added to the Furuta pendulum and the cartpole.

4.6 Related Work

In the related work section we embed the proposed algorithms within the existing literature. We summarize the existing work on continuous-time RL, value iteration and robust policy optimization and highlight the differences of our approach compared to prior art.

4.6.1 Continuous-Time Reinforcement Learning

The seminal work of Doya [21] introduced continuous-time RL. Since then, various approaches have been proposed to solve the Hamilton–Jacobi–Bellman (HJB) differential equation with the machine learning toolset. These methods can be divided into the trajectory and state-space-based methods. Trajectory-based methods solve the HJB along a trajectory to obtain the optimal trajectory. For example, path integral control uses the non-linear, control-affine dynamics with quadratic action costs to simplify the HJB to a linear partial differential equation [38, 62, 82, 83]. This differential equation can be transformed to a path-integral using the Feynman-Kac formulae. The path-integral can then be solved using Monte Carlo sampling to obtain the optimal state and action sequence. Recently, this approach has been used in combination with deep networks [64, 65, 70]. State-space-based methods solve the HJB globally to obtain an optimal non-linear controller applicable on the complete state domain. Classical approaches discretize the continuous spaces into a grid and solve the HJB or the robust Hamilton–Jacobi–Isaac (HJI) using a PDE solver [4]. In contrast, machine learning-based methods use function approximation and sampled states to solve the HJB. For example, regression-based approaches solved the HJB by fitting a radial-basis-function networks [21], deep networks [40, 41, 49, 79], kernels [34] or polynomial functions [48, 87] to minimize the HJB residual.

The presented work is closely related to the work of Kim et al. [40, 41]. While we present a model-based approach that leverages the dynamics model to perform value iteration, [40, 41] propose a model-free approach that uses Q-iteration. To obtain the Q-function, which usually does not exist for continuous-time RL [21], the authors incorporate a Lipschitz constraint that limits the change of the actions. Therefore, one does not control the actions but the change in the action. In this case, the optimal change in the action is the rescaled gradient of the Q-function w.r.t. the actions.

4.6.2 Fitted Value Iteration

Fitted Value Iteration (FVI) [3, 10, 58, 84] and the model-free counterpart fitted Q-Iteration [22, 23, 56, 73, 80] were previously only applicable to discrete actions and continuous states. These approaches were limited to discrete actions as they could not solve the maximum for continuous actions. Only QT-Opt [37] and NAF [29] applied fitted Q-Iteration to continuous actions. QT-Opt solves the maximization in each step using the particle-based cross-entropy method (CEM). However, this approach requires solving an expensive optimization problem within each step. In contrast, NAF uses a specific Q-function parametrization such that the Q-function is quadratic w.r.t. the actions. Therefore, the maximum can be easily computed due to the quadratic form. By leveraging the continuous-time formulation and the control-affine dynamics of many robotics systems, we showed that this maximization can be solved in closed form. Therefore, we extended FVI to continuous actions for high-frequency control tasks. For very low control frequencies this approximation might not be sufficient.

For continuous actions, current RL methods use policy iteration (PI) rather than VI [46, 74]. PI evaluates the value function of the current policy and hence, uses the action of the policy to compute the value function target. Therefore, PI circumvents the maximization required of VI (Eq. 4.20). In contrast to PI, our proposed method is 'policy-free' as the value function directly implies the policy. Therefore, cFVI and rFVI do not require the additional optimization to improve the policy as the PI-based methods.

4.6.3 Robust Policies for Sim2Real

Learning robust policies to bridge the simulation to reality gap has been approached by (1) changing the optimization objective [8, 9, 19], (2) using an adversary to optimize the worst-case performance [4, 32, 35, 36, 55, 57, 66, 67, 78, 81] and (3) randomizing the simulation [1, 16, 59, 59, 71, 85]. In this chapter, we focus on the adversarial formulation which has been used for continuous control tasks. For example, Pinto et al. [66, 67] used a separate agent as an adversary controlling an additive control input. This adversary maximized the negative reward using a standard deep RL algorithm. The agent and adversary do not share any information. Therefore, an additional optimization is required to optimize the adversary. Mandlekar et al. [55] used the auxiliary loss to maximize the policy actions. In contrast to these approaches, our approach is model-based instead of model-free. The model allows us to express the adversarial perturbations using analytic expressions derived directly from the

Hamilton–Jacobi–Isaacs (HJI) equation. Therefore, our approach shares knowledge between the actor and adversary due to a shared value function and requires no additional optimization.

Our proposed approach is similar to Morimoto and Doya [57]. In contrast to this work, we extend the analytic solutions to state, action, observation, and model disturbances, do not require a control-affine disturbance model, and use the constrained formulation rather than the penalized formulation.

4.7 Discussion

To obtain optimal and robust policies, we optimized the worst case reward rather than the expected reward, used dynamic programming on the complete state domain rather than local exploration and assumed a known dynamics model. While these assumptions enabled us to learn good policies for the sim2real transfer, these assumptions also have several drawbacks. We want to discuss the consequences of these assumptions and propose extensions to alleviate these limitations in future work.

4.7.1 *Worst Case Optimization*

The experiments showed that the worst-case optimization increases the policy robustness. However, the policy stiffness can also cause new problems. For example, the high stiffness of the policy makes the policy more susceptible to small time delays leading to chattering as observed on the Furuta Pendulum. Therefore, the worst-case optimization is a double-edged sword that depending on the system might be beneficial or cause additional problems. The admissible must be tuned manually to yield not overly conservative/pessimistic policies. One approach to overcome this limitation is to learn the magnitude of the admissible using data. One would interleave the offline planning with the online evaluation on the physical system. In every iteration, the obtained real-world data would be used to update the admissible set. Until one approaches a policy that can solve the task but is not overly conservative. Within the domain randomization community, this automatic tuning of the perturbed parameters has become widely used. Furthermore, one could parametrize the admissible set to be state-dependent to obtain higher robustness only when needed.

4.7.2 *State Distribution and Dimensionality*

The experiments showed that the state distribution significantly affects the policy robustness and performance. For the sim2real transfer, only the baselines with a uniform initial state distribution achieved a successful transfer to the physical system (Table 4.3). Furthermore, the dynamic programming variants, sampling uniformly from the complete state domain, performed much better than the real-time dynamic programming variants. This increased policy robustness is intuitive as the dynamic programming mitigates the distribution shift between simulation and the real-world system. However, the dynamic programming approaches cannot scale to high dimensional systems as sampling the complete state domain becomes unfeasible for such systems. For high-dimensional systems, the dataset $\mathcal{D}$ used to train DP cFVI and DP rFVI would be too sparse to learn an informative and consistent approximation of the value function. Therefore, the information about high reward states cannot be propagated to the relevant states. Gradient ascent on such value function would overfit to spurious solutions and would not solve the task. Therefore, an interesting future research question is how to obtain a sufficiently large state distribution such that the policy is robust when transferred to the physical system. This question is different from the traditional exploration exploitation trade-off as this question focuses more on minimizing the distribution mismatch.

4.7.3 Exploration

To improve the RTDP variants and scale to higher dimensional systems, the exploration of the proposed algorithms must be improved as the algorithms sometimes do not discover the optimal solution. This problem is especially pronounced for rFVI as in this case the adversary prevents discovering the optimal solution and the policy converges to a pessimistic policy (Fig. 4.5). The dynamic programming variants are not affected by this as this approach does not require exploration. The main problem for the exploration is the high-frequency sampling of the exploration noise required to solve the integrals. In this case, the exploration noise averages out and does not lead to diverse exploration. One approach to solve this would be to use model-predictive control for exploration. In this case, one would optimize the action sequence online and use the actions of the optimal policy only as prior.

4.7.4 Known Dynamics Model

All performed experiments used the analytic equations of motion provided by the manufacturer as the model. Therefore, we assumed that the model is known. The proposed policy optimization can be combined with model learning to learn the model from data. For example, a continuous-time control-affine system dynamics can be learned using Deep Lagrangian Networks [52, 53] or Hamiltonian Neural Network [28, 89]. In future work, one should combine continuous-time policy optimization with continuous-time model learning.

4.8 Summary

We have proposed continuous fitted value iteration (cFVI) and robust fitted value iteration (rFVI). These algorithms can be used to solve the Hamilton–Jacobi–Bellman (HJB) differential equation and the Hamilton–Jacobi–Isaacs (HJI) equation for continuous states and action spaces without grid-based samples. To derive these algorithms, we extended the existing derivations [21, 54, 57] of the optimal policy to a wider class of reward functions and introduced the solution for the optimal adversary. Instead of solving these equations directly using the regression techniques of machine learning as prior methods [21, 49, 79], we used value iteration to obtain a more reliable optimization. Thereby, we also extended fitted value iteration to continuous actions and adversarial RL. Previously, fitted value iteration was mainly applicable to discrete actions. The continuous control experiments showed that both algorithms can obtain the optimal policy and obtain identical reward as the deep reinforcement learning methods. Furthermore, the policies can be transferred to the physical systems. The rFVI policies are more robust when transferred to the physical system by applying higher actions. In addition, we provided an extensive discussion of the shortcomings and proposed approaches for future work to address these limitations. In addition to the quantitative comparison, a qualitative comparison of the learned value functions obtained by the different algorithms is performed (Fig. 4.1).

4.9 Appendix

4.9.1 Optimal Policy Proof—Theorem 4.1

Theorem *If the dynamics are control affine (Eq. 4.6), the reward is separable w.r.t. to state and action (Eq. 4.5) and the action cost g_c is positive definite and strictly convex, the continuous time optimal policy π^k w.r.t. V^k is described by*

$$\pi^*(\boldsymbol{x}) = \nabla \tilde{g}_c \left(\boldsymbol{B}(\boldsymbol{x})^\top \nabla_x V^*\right)$$

with the convex conjugate $\tilde{g}$ of g and the Jacobian of V w.r.t. the system state $\nabla_x V$.

Proof This proof follows the derivation our prior work [49] that generalized the special case described by [21]. Starting from the HJB equations and substituting the control affine dynamics yields

$$\begin{aligned}
\rho\, V^*(\boldsymbol{x}) &= \max_{\boldsymbol{u}}\ r(\boldsymbol{x}, \boldsymbol{u})\ + f_c(\boldsymbol{x}, \boldsymbol{u})^\top\, \nabla_x V^*, \\
&= \max_{\boldsymbol{u}}\ q_c(\boldsymbol{x}) - g_c(\boldsymbol{u})\ + [\boldsymbol{a}(\boldsymbol{x}) + \boldsymbol{B}(\boldsymbol{x})\boldsymbol{u}]\ \nabla_x V^* \\
&= \max_{\boldsymbol{u}}\ \left[\boldsymbol{u}^\top \boldsymbol{B}^\top \nabla_x V^* - g_c(\boldsymbol{u})\right] + q_c(\boldsymbol{x}) + \boldsymbol{a}^\top\, \nabla_x V^*.
\end{aligned}$$

Therefore, the continuous time optimal action is defined as

$$u^* = \arg\max_{u} \; u^\top B^\top \nabla_x V^* - g_c(u).$$

This optimization can be solved analytically as g_c is strictly convex and hence $\nabla g_c(u) = w$ is invertible, i.e., $u = [\nabla g_c]^{-1}(w) := \nabla \tilde{g}_c(w)$ with the convex conjugate $\tilde{g}$. The optimal action is described by

$$B(x)^\top \nabla_x V^* - \nabla g_c(u) := 0 \qquad \Rightarrow \qquad u^* = \nabla \tilde{g}_c\left(B(x)^\top \nabla_x V^*\right).$$

Therefore, the optimal action follows the gradient of the Value function upwards by projecting the gradient on using the control matrix and rescaling the magnitude using the action cost.

4.9.2 Optimal Adversary Proof—Theorem 4.2

4.9.2.1 State Disturbance Proof

Theorem *For the adversarial state disturbance (Eq. 4.10) with bounded in signal energy (Eq. 4.17), the optimal continuous time policy π^k and state disturbance ξ_x^k is described by*

$$\pi^*(x) = \nabla \tilde{g}\left(B(x)^\top \nabla_x V^*\right) \qquad \xi_x^* = -\alpha \frac{\nabla_x V^*}{\|\nabla_x V^*\|_2}.$$

Proof Equation 4.9 can be formulated with the explicit constraint instead of the infimum. The constrained HJI equation with an bounded signal energy constraint is described by

$$\begin{aligned} \rho\, V^*(x) = &\max_{u} \min_{\xi_x} \; r(x, u) \; + f_c(x, u, \xi_x)^\top \, \nabla_x V^*, \\ &\text{with } \xi_x^\top \xi_x - \alpha^2 \leq 0. \end{aligned}$$

Substituting the dynamics (Eq. 4.10) model and the reward yields

$$\begin{aligned} \rho V^* &= \max_{u} \min_{\xi_x} \; q_c(x) - g_c(u) \; + \left[a + Bu + \xi^*\right] \nabla_x V^*, \\ &= \max_{u} \left[u^\top B^\top \nabla_x V^* - g_c(u)\right] + \min_{\xi_x} \left[\xi_x^\top \nabla_x V^*\right] \\ &\qquad + \left[q_c + a^\top \nabla_x V^*\right]. \end{aligned}$$

Therefore, the optimal action identical to the optimal policy of the HJB and is described by

$$\boldsymbol{u}_t = \arg\max_{\boldsymbol{u}} \ \nabla_x V^\top \boldsymbol{B}(\boldsymbol{x}_t)\,\boldsymbol{u} - g_c(\boldsymbol{u})$$
$$\Rightarrow \quad \boldsymbol{u}_t = \nabla \tilde{g}_c\left(\boldsymbol{B}(\boldsymbol{x}_t)\nabla_x V\right).$$

The optimal state disturbance is described by

$$\boldsymbol{\xi}_x^* = \arg\min_{\boldsymbol{\xi}_x} \ \nabla_x V^\top \boldsymbol{\xi}_x \quad \text{with} \quad \boldsymbol{\xi}_x^\top \boldsymbol{\xi}_x - \alpha^2 \leq 0.$$

This constrained optimization can be solved using the Karush–Kuhn–Tucker (KKT) conditions, i.e.,

$$\nabla_x V + 2\lambda\, \boldsymbol{\xi}_x = 0 \quad \Rightarrow \quad \boldsymbol{\xi}_x^* = -\frac{1}{2\lambda}\nabla_x V$$

with the Lagrangian multiplier $\lambda \geq 0$. From primal feasibility and the complementary slackness condition of the KKT conditions follows that

$$\frac{1}{4\lambda^2}\nabla_x V^\top \nabla_x V - \alpha^2 \leq 0$$
$$\Rightarrow \lambda \geq \frac{1}{2\alpha}\sqrt{\nabla_x V^\top \nabla_x V}$$
$$\lambda\left(\frac{1}{4\lambda^2}\nabla_x V^\top \nabla_x V - \alpha^2\right) = 0$$
$$\Rightarrow \lambda_0 = 0, \ \ \lambda_1 = \frac{1}{2\alpha}\sqrt{\nabla_x V^\top \nabla_x V}.$$

Therefore, the optimal adversarial state perturbation with bounded signal energy is described by

$$\boldsymbol{\xi}^* = -\alpha \frac{V_x^*}{\|V_x^*\|_2}.$$

This solution is intuitive as the adversary wants to minimize the reward and this disturbance performs steepest descent using the largest step-size. Therefore, the perturbation is always on the constraint. □

4.9.2.2 Action Disturbance Proof

Theorem *For the adversarial action disturbance (Eq. 4.10) with bounded in signal energy (Eq. 4.17), the optimal continuous time policy π^k and action disturbance $\boldsymbol{\xi}_u^k$ is described by*

$$\pi^*(\boldsymbol{x}) = \nabla\tilde{g}\left(\boldsymbol{B}(\boldsymbol{x})^\top \nabla_x V^*\right) \qquad\qquad \boldsymbol{\xi}_u^* = -\alpha\frac{\boldsymbol{B}(\boldsymbol{x})^\top \nabla_x V^*}{\|\boldsymbol{B}(\boldsymbol{x})^\top \nabla_x V^*\|_2}.$$

Proof The HJI described in Eq. 4.9 can be formulated with the explicit constraint, i.e.,

$$\rho\, V^*(\boldsymbol{x}) = \max_{\boldsymbol{u}} \min_{\boldsymbol{\xi}_u}\ r(\boldsymbol{x}, \boldsymbol{u}) \ + f_c(\boldsymbol{x}, \boldsymbol{u}, \boldsymbol{\xi}_u)^\top\ \nabla_x V^*,$$
$$\text{with}\ \ \boldsymbol{\xi}_u^\top \boldsymbol{\xi}_u\ \leq \alpha^2.$$

Substituting the dynamics (Eq. 4.10) model and the reward yields

$$\begin{aligned}\rho V^* &= \max_{\boldsymbol{u}} \min_{\boldsymbol{\xi}_x}\ q_c(\boldsymbol{x}) - g_c(\boldsymbol{u}) \ + \left[\boldsymbol{a} + \boldsymbol{B}\,(\boldsymbol{u} + \boldsymbol{\xi})\right]\ \nabla_x V^*,\\ &= \max_{\boldsymbol{u}} \left[\boldsymbol{u}^\top \boldsymbol{B}^\top \nabla_x V^* - g_c(\boldsymbol{u})\right] + \min_{\boldsymbol{\xi}_x}\left[\boldsymbol{\xi}_u^\top \boldsymbol{B}^\top \nabla_x V^*\right]\\ &\qquad + \left[q_c + \boldsymbol{a}^\top \nabla_x V^*\right].\end{aligned}$$

Therefore, the optimal action is described by

$$\begin{aligned}\boldsymbol{u} &= \arg\max_{\boldsymbol{u}}\ \nabla_x V^\top \boldsymbol{B}(\boldsymbol{x})\,\boldsymbol{u} - g_c(\boldsymbol{u})\\ &\Rightarrow\ \boldsymbol{u}^* = \nabla\tilde{g}_c\left(\boldsymbol{B}(\boldsymbol{x})\nabla_x V^*\right).\end{aligned}$$

The optimal state disturbance is described by

$$\boldsymbol{\xi}_u^* = \arg\min_{\boldsymbol{\xi}_u}\ \nabla_x V^\top \boldsymbol{B}(\boldsymbol{x})\,\boldsymbol{\xi}_u \quad \text{with} \quad \boldsymbol{\xi}_u^\top \boldsymbol{\xi}_u \leq \alpha^2.$$

This constrained optimization can be solved using the Karush–Kuhn–Tucker (KKT) conditions, i.e.,

$$\boldsymbol{B}(\boldsymbol{x}_t)^\top \nabla_x V + 2\lambda\,\boldsymbol{\xi}_u = 0 \quad\Rightarrow\quad \boldsymbol{\xi}_u^* = -\frac{1}{2\lambda}\boldsymbol{B}(\boldsymbol{x}_t)^\top \nabla_x V$$

with the Lagrangian multiplier $\lambda \geq 0$. From primal feasibility and the complementary slackness condition of the KKT conditions follows that

$$\begin{aligned}&\frac{1}{4\lambda^2}\nabla_x V^\top \boldsymbol{B}\boldsymbol{B}^\top \nabla_x V - \alpha^2 \leq 0\\ &\qquad\qquad\qquad \Rightarrow \lambda \geq \frac{1}{2\alpha}\sqrt{\nabla_x V^\top \boldsymbol{B}\boldsymbol{B}^\top \nabla_x V}\\ &\lambda\left(\frac{1}{4\lambda^2}\nabla_x V^\top \boldsymbol{B}\boldsymbol{B}^\top \nabla_x V - \alpha^2\right) = 0\\ &\qquad\qquad\qquad \Rightarrow \lambda_0 = 0,\ \ \lambda_1 = \frac{1}{2\alpha}\sqrt{\nabla_x V^\top \boldsymbol{B}\boldsymbol{B}^\top \nabla_x V}.\end{aligned}$$

Therefore, the optimal adversarial state perturbation is described by

$$\boldsymbol{\xi}_u^* = -\alpha \frac{\boldsymbol{B}(\boldsymbol{x}_t)^\top V_x^*}{\|\boldsymbol{B}(\boldsymbol{x}_t)^\top V_x^*\|_2}.$$

This solution is similar to the state adversary but in this case the disturbance must be projected to the action space using the control matrix $\boldsymbol{B}$. □

4.9.2.3 Model Disturbance Proof

Theorem *For the adversarial model disturbance (Eq. 4.13) with element-wise bounded amplitude (Eq. 4.18), smooth drift and control matrix (i.e., $\boldsymbol{a}, \boldsymbol{B} \in C^1$) and $\boldsymbol{B}(\theta + \boldsymbol{\xi}_\theta) \approx \boldsymbol{B}(\theta)$, the optimal continuous time policy π^k and model disturbance $\boldsymbol{\xi}_\theta^k$ is described by*

$$\pi^*(\boldsymbol{x}) = \nabla \tilde{g}\left(\boldsymbol{B}(\boldsymbol{x})^\top \nabla_x V^*\right) \qquad \boldsymbol{\xi}_\theta^* = -\boldsymbol{\Delta}_\nu sign\left(z_\theta\right) + \boldsymbol{\mu}_\nu$$

$$\text{with } z_\theta = \left(\frac{\partial \boldsymbol{B}}{\partial \theta}\boldsymbol{u}^* + \frac{\partial \boldsymbol{a}}{\partial \theta}\right)^\top V_x^*,$$

mean $\boldsymbol{\mu}_\nu = \left(\boldsymbol{\nu}_{max} + \boldsymbol{\nu}_{min}\right)/2$ and range $\boldsymbol{\Delta}_\nu = \left(\boldsymbol{\nu}_{max} - \boldsymbol{\nu}_{min}\right)/2$.

Proof The HJI described in Eq. 4.9 can be written with the explicit constraint instead of the infimum with the admissible set Ω_A. With the bounded amplitude constraint, this constrained optimization is described by

$$\rho\, V^*(\boldsymbol{x}) = \max_{\boldsymbol{u}} \min_{\xi_x}\ r(\boldsymbol{x}, \boldsymbol{u}) \ + f_c(\boldsymbol{x}, \boldsymbol{u}, \boldsymbol{\xi}_\theta)^\top \nabla_{\boldsymbol{x}} V^*,$$

$$\text{with } \frac{1}{2}\left((\boldsymbol{\xi}_\theta - \boldsymbol{\mu}_\nu)^2 - \boldsymbol{\Delta}_\nu^2\right) \leq \boldsymbol{0}.$$

Substituting the dynamics (Eq. 4.13), the reward function as well as abbreviating $\boldsymbol{B}(\boldsymbol{x}; \theta + \boldsymbol{\xi}_\theta) = \boldsymbol{B}_\xi$ and $\boldsymbol{a}(\boldsymbol{x}; \theta + \boldsymbol{\xi}_\theta) = \boldsymbol{a}_\xi$ yields

$$\rho V^* = \max_{\boldsymbol{u}} \min_{\xi_\xi}\ q_c(\boldsymbol{x}) - g_c(\boldsymbol{u}) \ + \left[\boldsymbol{a}_\xi + \boldsymbol{B}_\xi \boldsymbol{u}\right]^\top \nabla_x V^*,$$

$$= \max_{\boldsymbol{u}} \min_{\xi_x} \left[\left[\boldsymbol{a}_\xi + \boldsymbol{B}_\xi \boldsymbol{u}\right]^\top \nabla_x V^* - g_c(\boldsymbol{u})\right] + q_c.$$

Therefore, the action and disturbance is determined by

$$\boldsymbol{u}^*, \boldsymbol{\xi}_\theta^* = \arg\max_{\boldsymbol{u}} \arg\min_{\xi} \left[\nabla_x V^\top \left(\boldsymbol{a}_\xi + \boldsymbol{B}_\xi \boldsymbol{u}\right) - g_c(\boldsymbol{u})\right].$$

This nested max-min optimization can be solved by first solving the inner optimization w.r.t. to $\boldsymbol{u}$ and substituting this solution into the outer maximization. The Lagrangian for the optimal model disturbance is described by

$$\boldsymbol{\xi}^* = \arg\min_{\boldsymbol{\xi}} \nabla_x V^\top \left(\boldsymbol{a}_\xi + \boldsymbol{B}_\xi \boldsymbol{u}\right) + \frac{1}{2}\boldsymbol{\lambda}^\top \left((\boldsymbol{\xi}_\theta - \boldsymbol{\mu}_\nu)^2 - \boldsymbol{\Delta}_\nu^2\right)$$

Using the KKT conditions this optimization can be solved. The stationarity condition yields

$$z_\theta + \boldsymbol{\lambda}^\top(\boldsymbol{\xi}_\theta - \boldsymbol{\mu}_\nu) = 0 \quad \Rightarrow \quad \boldsymbol{\xi}_\theta^* = -z_\theta \oslash \boldsymbol{\lambda} + \boldsymbol{\mu}_\nu$$

$$\text{with } z_\theta = \left[\frac{\partial \boldsymbol{a}}{\partial \theta} + \frac{\partial \boldsymbol{B}}{\partial \theta}\boldsymbol{u}\right]^\top \nabla_x V$$

and the element wise division $\oslash$. The primal feasibility and the complementary slackness yields

$$\frac{1}{2}\left(-z_\theta^2 \oslash \boldsymbol{\lambda}^2 - \boldsymbol{\Delta}_\nu^2\right) \leq 0 \quad \Rightarrow \quad \boldsymbol{\lambda} \geq \|z_\theta\|_1 \oslash \boldsymbol{\Delta}_\nu$$

$$\frac{1}{2}\boldsymbol{\lambda}^\top\left(-z_\theta^2 \oslash \boldsymbol{\lambda}^2 - \boldsymbol{\Delta}_\nu^2\right) = 0 \quad \Rightarrow \quad \boldsymbol{\lambda}_0 = \boldsymbol{0}, \;\; \boldsymbol{\lambda}_1 = \|z_\theta\|_1 \oslash \boldsymbol{\Delta}_\nu.$$

Therefore, the optimal model disturbance is described by

$$\boldsymbol{\xi}_\theta^*(\boldsymbol{u}) = -\boldsymbol{\Delta}_\nu sign\big(z_\theta(\boldsymbol{u})\big) + \boldsymbol{\mu}_\nu$$

as $z_\theta \oslash \|z_\theta\|_1 = sign(z_\theta)$. Then the optimal action can be computed by

$$\boldsymbol{u}^* = \arg\max_{\boldsymbol{u}} \nabla_x V^\top \left[\boldsymbol{a}(\boldsymbol{\xi}_\theta^*(\boldsymbol{u})) + \boldsymbol{B}\left(\boldsymbol{\xi}_\theta^*(\boldsymbol{u})\right)\boldsymbol{u}\right] - g_c(\boldsymbol{u}).$$

Due to the envelope theorem, the extrema is described by

$$\boldsymbol{B}(\boldsymbol{x}; \theta + \boldsymbol{\xi}^*(\boldsymbol{u}))^\top \nabla_x V - g_c(\boldsymbol{u}) = 0.$$

This expression cannot be solved without approximation as $\boldsymbol{B}$ does not necessarily be invertible w.r.t. θ. Approximating $\boldsymbol{B}(\boldsymbol{x}; \theta + \boldsymbol{\xi}^*(\boldsymbol{u})) \approx \boldsymbol{B}(\boldsymbol{x}; \theta)$, lets one solve for $\boldsymbol{u}$. In this case the optimal action $\boldsymbol{u}^*$ is described by $\boldsymbol{u}^* = \nabla\tilde{g}(\boldsymbol{B}(\boldsymbol{x}; \theta)^\top \nabla_x V)$. This approximation is feasible for two reasons. First of all, if the adversary can significantly alter the dynamics in each step, the system would not be controllable and the optimal policy would not be able to solve the task. Second, this approximation implies that neither agent or the adversary can react to the action of the other and must choose simultaneously. This assumption is common in prior works [4]. The order of the minimization and maximization is interchangeable. For both cases the optimal action as well as optimal model disturbance are identical and require the same approximation during the derivation. □

4.9.2.4 Observation Disturbance Proof

Theorem *For the adversarial observation disturbance (Eq. 4.12) with bounded signal energy (Eq. 4.17), smooth drift and control matrix (i.e., $\boldsymbol{a}, \boldsymbol{B} \in C^1$) and $\boldsymbol{B}(\boldsymbol{x} + \boldsymbol{\xi}_o) \approx \boldsymbol{B}(\boldsymbol{x})$, the optimal continuous time policy π^k and observation disturbance $\boldsymbol{\xi}_o^k$ is described by*

$$\pi^*(\boldsymbol{x}) = \nabla \tilde{g} \left(\boldsymbol{B}(\boldsymbol{x})^\top \nabla_x V^* \right) \qquad \boldsymbol{\xi}_o^* = -\alpha \frac{z_o}{\|z_o\|_2}$$

$$\textit{with } z_o = \left(\frac{\partial \boldsymbol{a}(\boldsymbol{x};\, \theta)}{\partial \boldsymbol{x}} + \frac{\partial \boldsymbol{B}(\boldsymbol{x};\, \theta)}{\partial \boldsymbol{x}} \boldsymbol{u}^* \right)^\top V_x^*.$$

Proof The HJI (Eq. 4.9) can be written with the explicit constraint instead of the infimum with the admissible set Ω_A. The constrained optimization with an bounded energy constraint is described by

$$\rho\, V^*(\boldsymbol{x}) = \max_{\boldsymbol{u}} \min_{\boldsymbol{\xi}_x} \; r(\boldsymbol{x}, \boldsymbol{u}) \; + f_c(\boldsymbol{x}, \boldsymbol{u}, \boldsymbol{\xi}_o)^\top \, \nabla_x V^*,$$
$$\text{with } \frac{1}{2} \left(\boldsymbol{\xi}_o^\top \boldsymbol{\xi}_o - \alpha^2 \right) \le \boldsymbol{0}.$$

Substituting the dynamics model (Eq. 4.12), the reward function as well as abbreviating $\boldsymbol{B}(\boldsymbol{x} + \boldsymbol{\xi}_o; \theta)$ as $\boldsymbol{B}_\xi$ yields

$$\begin{aligned}
\rho V^* &= \max_{\boldsymbol{u}} \min_{\boldsymbol{\xi}_\xi} \; q_c(\boldsymbol{x}) - g_c(\boldsymbol{u}) \; + \left[\boldsymbol{a}_\xi + \boldsymbol{B}_\xi \boldsymbol{u} \right]^\top \, \nabla_x V^*, \\
&= \max_{\boldsymbol{u}} \min_{\boldsymbol{\xi}_x} \left[\left[\boldsymbol{a}_\xi + \boldsymbol{B}_\xi \boldsymbol{u} \right]^\top \, \nabla_x V^* - g_c(\boldsymbol{u}) \right] + q_c.
\end{aligned}$$

Similar to the model disturbance, this nested max-min optimization can be solved by first solving the inner optimization w.r.t. to ξ and substituting this solution into the outer maximization. The Lagrangian for the optimal model disturbance is described by

$$\boldsymbol{\xi}_o^* = \arg\min_{\boldsymbol{\xi}_o} \; \nabla_x V^\top \left(\boldsymbol{a}(\boldsymbol{\xi}_o) + \boldsymbol{B}(\boldsymbol{\xi}_o) \boldsymbol{u} \right) + \frac{\lambda}{2} \left(\boldsymbol{\xi}_o^\top \boldsymbol{\xi}_o - \alpha^2 \right)$$

Using the KKT conditions this optimization can be solved. The stationarity condition yields

$$z_o + \lambda\, \boldsymbol{\xi}_o = 0 \quad \Rightarrow \quad \boldsymbol{\xi}_o^* = -\frac{1}{\lambda} z_o$$

$$\text{with } z_o = \left[\frac{\partial \boldsymbol{a}(\boldsymbol{x};\, \theta)}{\partial \boldsymbol{x}} + \frac{\partial \boldsymbol{B}(\boldsymbol{x};\, \theta)}{\partial \boldsymbol{x}} \boldsymbol{u}^*\right]^\top \nabla_x V.$$

The primal feasibility and the complementary slackness yield

$$\frac{1}{2}\left(\frac{1}{\lambda^2} z_o^\top z_o - \alpha^2\right) \leq 0 \quad \Rightarrow \quad \lambda \geq \frac{1}{\alpha} \|z_\theta\|_2$$

$$\frac{\lambda}{2}\left(\frac{1}{\lambda^2} z_o^\top z_o - \alpha^2\right) = 0 \quad \Rightarrow \quad \lambda_0 = 0,\ \ \lambda_1 = \frac{1}{\alpha} \|z_o\|_2.$$

Therefore, the optimal observation disturbance is described by

$$\boldsymbol{\xi}_o^*(\boldsymbol{u}) = -\alpha \frac{z_o}{\|z_o\|_2}.$$

Then the optimal action can be computed by

$$\boldsymbol{u}^* = \arg\max_{\boldsymbol{u}} \nabla_x V^\top \left[\boldsymbol{a}\left(\boldsymbol{\xi}_o^*(\boldsymbol{u})\right) + \boldsymbol{B}\left(\boldsymbol{\xi}_o^*(\boldsymbol{u})\right) \boldsymbol{u}\right] - g_c(\boldsymbol{u}).$$

Due to the envelope theorem, the extrema is described by

$$\boldsymbol{B}(\boldsymbol{x}; \theta + \boldsymbol{\xi}_o^*(\boldsymbol{u}))^\top \nabla_x V - g_c(\boldsymbol{u}) = 0.$$

This expression cannot be solved without approximation as $\boldsymbol{B}$ does not necessarily be invertible w.r.t. $\boldsymbol{x}$. Approximating $\boldsymbol{B}(\boldsymbol{x} + \boldsymbol{\xi}_o^*(\boldsymbol{u});\, \theta) \approx \boldsymbol{B}(\boldsymbol{x}; \theta)$, lets one solve for $\boldsymbol{u}$. In this case the optimal action $\boldsymbol{u}^*$ is described by $\boldsymbol{u}^* = \nabla \tilde{g}(\boldsymbol{B}(\boldsymbol{x}; \theta)^\top \nabla_x V)$. This approximation is feasible for two reasons. First of all, if the adversary can significantly alter the dynamics in each step, the system would not be controllable and the optimal policy would not be able to solve the task. Second, this approximation implies that neither agent or the adversary can react to the action of the other and must choose simultaneously. This assumption is common in prior works [4]. The order of the minimization and maximization is interchangeable. For both cases the optimal action as well as optimal model disturbance are identical and require the same approximation during the derivation. □

References

1. Andrychowicz, O.M., Baker, B., Chociej, M., Jozefowicz, R., McGrew, B., Pachocki, J., Petron, A., Plappert, M., Powell, G., Ray, A., et al.: Learning dexterous in-hand manipulation. Int. J. Robot. Res. (2020)
2. Aström, K.J.: Event based control. In: Analysis and Design of Nonlinear Control Systems. Springer (2008)
3. Baird, L.: Residual algorithms: reinforcement learning with function approximation. In: Machine Learning Proceedings. Elsevier (1995)
4. Bansal, S., Chen, M., Herbert, S., Tomlin, C.J.: Hamilton-Jacobi reachability: a brief overview and recent advances. In: Conference on Decision and Control (CDC) (2017)
5. Barto, A.G., Bradtke, S.J., Singh, S.P.: Learning to act using real-time dynamic programming. Artif. Intell. (1995)
6. Bellman, R.: Dynamic Programming. Princeton University Press (1957)
7. Berkenkamp, F., Turchetta, M., Schoellig, A., Krause, A.: Safe model-based reinforcement learning with stability guarantees. In: Advances in Neural Information Processing Systems (2017)
8. Bharadhwaj, H., Kumar, A., Rhinehart, N., Levine, S., Shkurti, F., Garg, A.: Conservative safety critics for exploration. In: International Conference on Learning Representations (ICLR) (2021)
9. Borkar, V.S.: A sensitivity formula for risk-sensitive cost and the actor–critic algorithm. Syst. & Control Lett. (2001)
10. Boyan, J., Moore, A.: Generalization in reinforcement learning: safely approximating the value function. In: Advances in Neural Information Processing Systems (NeurIPS) (1994)
11. Boyd, S., Vandenberghe, L.: Convex Optimization. Cambridge University Press (2004)
12. Buckman, J., Hafner, D., Tucker, G., Brevdo, E., Lee, H.: Sample-efficient reinforcement learning with stochastic ensemble value expansion. In: Advances in Neural Information Processing Systems (NeurIPS) (2018)
13. Caputo, M.R., Caputo, M.R.: Foundations of Dynamic Economic Analysis: Optimal Control Theory and Applications. Cambridge University Press (2005)
14. Carter, M.: Foundations of Mathematical Economics. MIT Press (2001)
15. Chang, Y.C., Roohi, N., Gao, S.: Neural lyapunov control. In: Advances in Neural Information Processing Systems (NeurIPS) (2019)
16. Chebotar, Y., Handa, A., Makoviychuk, V., Macklin, M., Issac, J., Ratliff, N., Fox, D.: Closing the sim-to-real loop: adapting simulation randomization with real world experience. In: International Conference on Robotics and Automation (ICRA) (2019)
17. Chen, M., Fisac, J.F., Sastry, S., Tomlin, C.J.: Safe sequential path planning of multi-vehicle systems via double-obstacle Hamilton-Jacobi-Isaacs variational inequality. In: European Control Conference (ECC) (2015)
18. Ching, S., Eun, Y., Gokcek, C., Kabamba, P.T., Meerkov, S.M.: Quasilinear Control: Performance Analysis and Design of Feedback Systems with Nonlinear Sensors and Actuators. Cambridge University Press (2010)
19. Chow, Y., Tamar, A., Mannor, S., Pavone, M.: Risk-sensitive and robust decision-making: a cvar optimization approach (2015). arXiv:1506.02188
20. De Doná, J.A., Goodwin, G.C.: Elucidation of the state-space regions wherein model predictive control and anti-windup strategies achieve identical control policies. In: American Control Conference (ACC). IEEE (2000)
21. Doya, K.: Reinforcement learning in continuous time and space. Neural Comput. (2000)
22. Ernst, D., Geurts, P., Wehenkel, L.: Tree-based batch mode reinforcement learning. J. Mach. Learn. Res. (2005)
23. Massoud Farahmand, A., Ghavamzadeh, M., Szepesvári, C., Mannor, S.: Regularized fitted q-iteration for planning in continuous-space markovian decision problems. In: American Control Conference (ACC). IEEE (2009)
24. Feinberg, V., Wan, A., Stoica, I., Jordan, M.I., Gonzalez, J.E., Levine, S.: Model-based value estimation for efficient model-free reinforcement learning (2018). arXiv:1803.00101

25. Fisac, J.F., Akametalu, A.K., Zeilinger, M.N., Kaynama, S., Gillula, J., Tomlin, C.J.: A general safety framework for learning-based control in uncertain robotic systems. IEEE Trans. Autom. Control (2018)
26. Fleming, W.H., Soner, H.M.: Controlled Markov processes and viscosity solutions. Springer Science & Business Media (2006)
27. Gleave, A., Dennis, M., Wild, C., Kant, N., Levine, S., Russell, S.: Adversarial policies: attacking deep reinforcement learning (2019). arXiv:1905.10615
28. Greydanus, S., Dzamba, M., Yosinski, J.: Hamiltonian neural networks. In: Advances in Neural Information Processing Systems (NeuRIPS) (2019)
29. Gu, S., Lillicrap, T., Sutskever, I., Levine, S.: Continuous deep q-learning with model-based acceleration. In: International Conference on Machine Learning (ICML) (2016)
30. Gulcehre, C., Wang, Z., Novikov, A., Paine, T.L., Colmenarejo, S.G., Zolna, K., Agarwal, R., Merel, J., Mankowitz, D., Paduraru, C., et al.: Rl unplugged: Benchmarks for offline reinforcement learning (2020). arXiv:2006.13888
31. Haarnoja, T., Zhou, A., Abbeel, P., Levine, S.: Soft actor-critic: off-policy maximum entropy deep reinforcement learning with a stochastic actor. In: International Conference on Machine Learning (ICML) (2018)
32. Harrison*, J., Garg*, A., Ivanovic, B., Zhu, Y., Savarese, S., Fei-Fei, L., Pavone (* equal contribution), M.: AdaPT: zero-shot adaptive policy transfer for stochastic dynamical systems. In: International Symposium on Robotics Research (ISRR). Springer STAR (2017)
33. Heger, M.: Consideration of risk in reinforcement learning. In: Machine Learning Proceedings. Elsevier (1994)
34. Hennig, P.: Optimal reinforcement learning for gaussian systems. In: Advances in Neural Information Processing Systems (2011)
35. Hsu, K.C., Rubies-Royo, V., Tomlin, C.J., Fisac, J.F.: Safety and liveness guarantees through reach-avoid reinforcement learning. Robot.: Sci. Syst. (2021)
36. Isaacs, R.: Differential Games: A Mathematical Theory with Applications to Warfare and Pursuit, Control and Optimization. Courier Corporation (1999)
37. Kalashnikov, D., Irpan, A., Pastor, P., Ibarz, J., Herzog, A., Jang, E., Quillen, D., Holly, E., Kalakrishnan, M., Vanhoucke, V., et al.: Qt-opt: scalable deep reinforcement learning for vision-based robotic manipulation (2018). arXiv:1806.10293
38. Kappen, H.J.: Linear theory for control of nonlinear stochastic systems. Phys. Rev. Lett. (2005)
39. Khalil, H.K., Grizzle, J.W.: Nonlinear Systems. Prentice Hall, Upper Saddle River, NJ (2002)
40. Kim, J., Shin, J., Yang, I.: Hamilton-Jacobi deep Q-Learning for deterministic continuous-time systems with lipschitz continuous controls (2020). arXiv:2010.14087
41. Kim, J., Yang, I.: Hamilton-Jacobi-Bellman equations for Q-Learning in continuous time. In: Learning for Dynamics and Control (2020)
42. Kirk, D.E.: Optimal Control Theory: An Introduction. Courier Corporation (1970)
43. Kolter, J.Z., Manek, G.: Learning stable deep dynamics models. In: Advances in Neural Information Processing Systems (NeurIPS) (2019)
44. LaValle, S.M.: Planning Algorithms. Cambridge University Press (2006)
45. Liberzon, D.: Calculus of variations and optimal control theory: a concise introduction. Princeton University Press (2011)
46. Lillicrap, T.P., Hunt, J.J., Pritzel, A., Heess, N., Erez, T., Tassa, Y., Silver, D., Wierstra, D.: Continuous control with deep reinforcement learning (2015). arXiv:1509.02971
47. Littman, M.L.: Markov games as a framework for multi-agent reinforcement learning. In: Machine Learning Proceedings. Elsevier (1994)
48. Liu, D., Wang, D., Wang, F.Y., Li, H., Yang, X.: Neural-network-based online hjb solution for optimal robust guaranteed cost control of continuous-time uncertain nonlinear systems. IEEE Trans. Cybern. (2014)
49. Lutter, M., Belousov, B., Listmann, K., Clever, D., Peters, J.: HJB optimal feedback control with deep differential value functions and action constraints. In: Conference on Robot Learning (CoRL) (2019)

50. Lutter, M., Mannor, S., Peters, J., Fox, D., Garg, A.: Robust value iteration for continuous control tasks. In: Robotics: Science and Systems (RSS) (2021)
51. Lutter, M., Mannor, S., Peters, J., Fox, D., Garg, A.: Value iteration in continuous actions, states and time. In: International Conference on Machine Learning (ICML) (2021)
52. Lutter, M., Peters, J.: Deep lagrangian networks for end-to-end learning of energy-based control for under-actuated systems. In: International Conference on Intelligent Robots and Systems (IROS) (2019)
53. Lutter, M., Ritter, C., Peters, J.: Deep lagrangian networks: using physics as model prior for deep learning. In: International Conference on Learning Representations (ICLR) (2019)
54. Lyshevski, S.E.: Optimal control of nonlinear continuous-time systems: design of bounded controllers via generalized nonquadratic functionals. In: American Control Conference (ACC). IEEE (1998)
55. Mandlekar, A., Zhu, Y., Garg, A., Fei-Fei, L., Savarese, S.: Adversarially robust policy learning: active construction of physically-plausible perturbations. In: International Conference on Intelligent Robots and Systems (IROS) (2017)
56. Mnih, V., Kavukcuoglu, K., Silver, D., Rusu, A.A., Veness, J., Bellemare, M.G., Graves, A., Riedmiller, M., Fidjeland, A.K., Ostrovski, G., et al.: Human-level control through deep reinforcement learning. Nature (2015)
57. Morimoto, J., Doya, K.: Robust reinforcement learning. Neural Comput. (2005)
58. Munos, R., Szepesvári, C.: Finite-time bounds for fitted value iteration. J. Mach. Learn. Res. (2008)
59. Muratore, F., Eilers, C., Gienger, M., Peters, J.: Data-efficient domain randomization with bayesian optimization. In: IEEE Robotics and Automation Letters (RAL) (2021)
60. Muratore, F., Treede, F., Gienger, M., Peters, J.: Domain randomization for simulation-based policy optimization with transferability assessment. In: Conference on Robot Learning (CoRL) (2018)
61. Nilim, A., El Ghaoui, L.: Robust control of markov decision processes with uncertain transition matrices. Oper. Res. (2005)
62. Pan, Y., Theodorou, E.A., Kontitsis, M.: Model-based path integral stochastic control: a bayesian nonparametric approach (2014). arXiv:1412.3038
63. Pattanaik, A., Tang, Z., Liu, S., Bommannan, G., Chowdhary, G.: Robust deep reinforcement learning with adversarial attacks (2017). arXiv:1712.03632
64. Pereira, M., Wang, Z., Exarchos, I., Theodorou, E.: Learning deep stochastic optimal control policies using forward-backward sdes. In: Robotics: Science and Systems (2019)
65. Pereira, M.A., Wang, Z., Exarchos, I., Theodorou, E.A.: Safe optimal control using stochastic barrier functions and deep forward-backward sdes (2020). arXiv:2009.01196
66. Pinto, L., Davidson, J., Gupta, A.: Supervision via competition: robot adversaries for learning tasks. In: International Conference on Robotics and Automation (ICRA) (2017)
67. Pinto, L., Davidson, J., Sukthankar, R., Gupta, A.: Robust adversarial reinforcement learning. In: International Conference on Machine Learning (ICML) (2017)
68. Puterman, M.L.: Markov Decision Processes: Discrete Stochastic Dynamic Programming. Wiley (1994)
69. Quanser: Quanser courseware and resources (2018). https://www.quanser.com/solution/control-systems/
70. Rajagopal, K., Balakrishnan, S.N., Busemeyer, J.R.: Neural network-based solutions for stochastic optimal control using path integrals. IEEE Trans. Neural Netw. Learn. Syst. (2016)
71. Ramos, F., Possas, R.C., Fox, D.: Bayessim: adaptive domain randomization via probabilistic inference for robotics simulators (2019). arXiv:1906.01728
72. Richards, S.M., Berkenkamp, F., Krause, A.: The lyapunov neural network: adaptive stability certification for safe learning of dynamical systems (2018). arXiv:1808.00924
73. Riedmiller, M.: Neural fitted q iteration–first experiences with a data efficient neural reinforcement learning method. In: European Conference on Machine Learning. Springer (2005)
74. Schulman, J., Moritz, P., Levine, S., Jordan, M., Abbeel, P.: High-dimensional continuous control using generalized advantage estimation (2015). arXiv:1506.02438

75. Schulman, J., Wolski, F., Dhariwal, P., Radford, A., Klimov, O.: Proximal policy optimization algorithms (2017). arXiv:1707.06347
76. Sutton, R.S., Barto, A.G., et al.: Introduction to Reinforcement Learning. MIT Press, Cambridge (1998)
77. Tallec, C., Blier, L., Ollivier, Y.: Making deep q-learning methods robust to time discretization (2019). arXiv:1901.09732
78. Tamar, A., Xu, H., Mannor, S.: Scaling up robust mdps by reinforcement learning (2013). arXiv:1306.6189
79. Tassa, Y., Erez, T.: Least squares solutions of the HJB equation with neural network value-function approximators. IEEE Trans. Neural Netw. (2007)
80. Tesauro, G.: Practical issues in temporal difference learning. Mach. Learn. (1992)
81. Tessler, C., Efroni, Y., Mannor, S.: Action robust reinforcement learning and applications in continuous control. In: International Conference on Machine Learning (ICML) (2019)
82. Theodorou, E., Buchli, J., Schaal, S.: Reinforcement learning of motor skills in high dimensions: a path integral approach. In: International Conference on Robotics and Automation (ICRA). IEEE (2010)
83. Todorov, E.: Linearly-solvable markov decision problems. In: Advances in Neural Information Processing Systems (2007)
84. Tsitsiklis, J.N., Van Roy, B.: Feature-based methods for large scale dynamic programming. Mach. Learn. (1996)
85. Xie, Z., Da, X., van de Panne, M., Babich, B., Garg, A.: Dynamics randomization revisited: a case study for quadrupedal locomotion. In: IEEE International Conference on Robotics and Automation (ICRA) (2021)
86. Xu, H., Mannor, S.: Robustness and generalization. Mach. Learn. (2012)
87. Yang, X., Liu, D., Wang, D.: Reinforcement learning for adaptive optimal control of unknown continuous-time nonlinear systems with input constraints. Int. J. Control (2014)
88. Zhang, H., Chen, H., Xiao, C., Li, B., Boning, D., Hsieh, C.J.: Robust deep reinforcement learning against adversarial perturbations on observations (2020). arXiv:2003.08938
89. Zhong, Y.D., Dey, B., Chakraborty, A.: Symplectic ode-net: learning hamiltonian dynamics with control (2019). arXiv:1909.12077

Chapter 5
Conclusion

In this book, we showed that one can combine inductive biases with machine learning to obtain learning algorithms suitable for robotics and control. In contrast to prior work, the proposed algorithms cover the complete spectrum between classical methods and black-box deep learning. Even for deep networks, one can leverage domain knowledge to obtain deep network representations that retain some of the benefits of classical robotics approaches. The main take-away of this book is that one can use deep networks in more creative ways than naive input-output mappings for learning dynamics models or policies. In the following, we summarize the contributions of the three chapters and discuss the open challenges of the presented algorithms.

5.1 Summary

We presented three algorithms that incorporate structure within the learning algorithms to improve the performance of robot control. Chapter 1 introduced the motivation for this book. We discussed the advantages and disadvantages of classical engineering and end-to-end learning with deep networks to program a robot. Subsequently, we stated the research question and the difference to the prior art.

Chapter 2 focused on learning dynamics models when the kinematic chain is known. We presented the Differentiable Newton–Euler Algorithm (DiffNEA) that can infer physically consistent simulator parameters for rigid body systems augmented with various friction models and systems with non-holonomic constraints. In contrast to the classical approach [2, 9, 16, 19], this approach leverages automatic differentiation, virtual parameters, and gradient-based optimization to infer parameters that are guaranteed to be physically plausible. Within the experimental evaluation, we showed that this approach excels when extrapolation is required. DiffNEA learned an accurate dynamics model of ball in a cup that includes the string and cup dynamics with only 4 min of data. When used for model-based

M. Lutter, *Inductive Biases in Machine Learning for Robotics and Control*,
Springer Tracts in Advanced Robotics 156,
https://doi.org/10.1007/978-3-031-37832-4_5

reinforcement learning, a policy was obtained that was transferable to the physical system. The black box deep networks were not able to solve the task. The reinforcement learning exploited the dynamics models and converged to random movements. Therefore, the DiffNEA model enables generalization beyond the training domain and is very data efficient due to the incorporated structure.

Chapter 3 focused on learning dynamics models of mechanical systems when no specific knowledge of the system is known. We presented Deep Lagrangian Networks (DeLaN) that combines deep networks with Lagrangian mechanics to learn dynamics models that conserve energy. DeLaN uses two deep networks to represent the potential and kinetic energy of the system. These networks are combined to approximate the Lagrangian. The forward and inverse model of the system can be computed using the approximated Lagrangian and the Euler–Lagrange differential equation. The system energy can be learned unsupervised by minimizing the squared residual of the Euler–Lagrange equation. The resulting DeLaN models retain many advantages of classical system identification techniques but do not require any specific knowledge of the individual system. DeLaN conserves energy, enables energy control, and is interpretable, i.e., can be used to compute Coriolis force, gravitational force, generalized momentum, and other physical quantities. Within the experimental evaluation, we highlighted that these learned models can be used for real-time control of simulated and physical rigid body systems. Compared to standard deep networks, the physics-inspired models learn better models and capture the underlying structure of the dynamics. The DeLaN models also enable energy control and successfully swing up the under-actuated Furuta pendulum and the cartpole. Previously, this energy control was not possible using black-box model learning approaches as these cannot learn the energy. Since our initial introduction of DeLaN many variants of physics-inspired deep networks have been proposed that use Hamiltonian dynamics, add friction models or use feature transformation [6, 11, 12, 23, 26].

Chapter 4 focused on learning robust optimal policies that bridge the simulation to reality gap. We presented Robust Fitted Value Iteration (rFVI) that learns a robust optimal policy by solving the adversarial reinforcement learning problem with value iteration. Leveraging the non-linear control-affine dynamics of many mechanical systems and the separable state and action reward of many continuous control problems, we derive the optimal policy and optimal adversary in closed form. These analytic expressions enable us to extend value iteration to continuous actions and states as well as solving the two-player zero-sum game. Notably, the resulting algorithms do not require discretization of states or actions. Within the experimental evaluation, we highlighted that these learned policies can control under-actuated systems in real-time and successfully achieve the simulation to reality transfer. When changing the masses of the pendulum, the robust rFVI policy performs better compared to deep reinforcement learning algorithm with uniform domain randomization.

In summary, this book presented three novel algorithms that can be used for learning models or obtaining control policies. Each algorithm leveraged domain knowledge and was evaluated on the physical system. Compared to the black box deep networks, the presented algorithms improve the performance and enabled new applications.

5.2 Open Problems and Future Work

The presented algorithms are an initial proof of concept of this combination and showcase the potential of data-driven learning with inductive biases. These algorithms do not solve all problems of black-box deep learning-based approaches or classical engineering approaches and have their limitations. In this section, we elaborate on the open problems of the presented methods and describe interesting directions of future research.

5.2.1 *Learning Dynamics Models*

We presented two algorithms to learn dynamics models, DiffNEA and DeLaN. While these methods can learn dynamics models, these methods can only be applied to systems without contact, require the observation of generalized coordinates, optimize the 1-step or multi-step mean squared error and require the manual encoding of conservation laws and symmetries.

Contacts. Learning the dynamics of contact-rich tasks is important as most interesting real-world problems include contacts. Many of the existing model learning approaches including the presented DiffNEA and DeLaN and the other physics-inspired networks only focus on articulated bodies that do not interact with the environment. Theoretically, contacts can be included in DeLaN or DiffNEA using an analytic contact model. In the general case, a collision checker determines all contact points and their respective Jacobians. Afterward, the contact force is computed by solving the linear complementarity problem. Within the literature, various researchers have shown that this computation is differentiable [3, 8, 10, 13, 15, 25]. Therefore, this contact model can be added to both presented approaches. Various researchers have also used this approach for trivial examples, e.g., bouncing discs, Newton cradle, and an n-link pendulum with floor [14, 27].

The main problem of this approach is the underlying assumption that the kinematic chain and meshes of the body are known. Both models are required by the collision checker to determine the contact point and Jacobian. While this assumption is tolerable for DiffNEA, for DeLaN this assumption is too restrictive. In this case, the requirements of DeLaN and DiffNEA are identical but DeLaN does not obtain a global model. For DiffNEA, future work should try to scale the combination of an analytical contact model with DiffNEA to multi-contact problems and use a learnable representation for the meshes. Therefore, not only the simulator parameters are learned but also the meshes of the links. A different way to include contacts within DeLaN would be to model the contacts as potential fields that push apart penetrated objects. The main challenge for this approach is to learn consistent potential fields in regions where no data is observed. As the training data does not contain penetrated objects, these potential fields cannot be learned supervised.

Generalized Coordinates. The second limiting assumption is the observation of the generalized coordinates, momentum, and forces. Most robotic systems that do not only involve a rigid body manipulator, but these coordinates are commonly not observed or known. One usually only obtains observations derived from the generalized coordinates if the system is fully observed. In many cases, the system is only partially observed and one cannot infer the system state from a single observation. For physics-inspired networks such as DeLaN, variational autoencoders (VAE) have been proposed to extend these methods to unstructured observations. In this case, the VAE is supposed to learn a latent space that resembles the generalized coordinates, and the Lagrangian and Hamiltonian dynamics are applied in the latent space. For artificial images and simple systems such as single-link pendulums and N-body problems, this approach has been successful [1, 11, 23, 24, 26]. However, for more complex systems and realistic rendering of systems this approach does not generate better models than traditional deep networks [4]. Future work needs to revisit approaches to learn latent spaces that retain the important information and symmetries.

Optimization Loss. The presented model learning approaches optimize the 1-step or multi-step mean squared prediction error (mse). This optimization loss has two problems. First, it has been shown that the 1-step and n-step mse does not correlate with a planning performance [17, 18]. Second, some parameters have negligible impact on the 1-step loss but detrimental impact on long-term prediction. For example, the friction of the ball in the ball in cup experiment is barely observable on the 1-step optimization loss but becomes important when considering the complete ball trajectory. A similar effect can be observed for the joint friction of the Furuta pendulum. Simply using the n-step loss for these problems is not sufficient as backprop through time can usually not be applied to such long horizons.

A different approach would be to optimize an adversarial loss instead of the prediction error. This approach would be similar to a generative adversarial network, where one learns a discriminator that differentiates between simulated and predicted trajectories. This approach could potentially alleviate both shortcomings of the prediction error. First, one uses an adversarial setting which should reduce the exploitability of the learned model. Hence, the performance measure should be a better proxy for predicting the planning performance. Second, the discriminator could learn a better distance measure between long trajectories than the naive mean squared error. Therefore, the discriminator should create better supervising feedback that captures the long-term impact of the different parameters. Furthermore, one could use network architectures from speech generation and recognition literature to include the temporal characteristics within the discriminator.

Conservation Law and Symmetries. The presented methods as well as related model learning methods hard-code the conservation laws and symmetries within the losses or model architecture. However, ultimately one wants to discover the invariances and equivariances from data. Initial works have combined deep learning with symbolic regression to infer the physical laws [7]. However, the current methods only work for simple relations. Therefore, a promising and under-explored research direction would be to infer the symmetries and conservation laws from data. Furthermore, this approach would also extend to learning optimal control policies as these policies commonly have symmetries and current deep network approaches do not identify or leverage these symmetries.

5.2.2 *Learning Robust Policies*

The presented dynamic programming approach rFVI can compute the optimal policy that is robust to changes in the dynamics. However, this approach has two disadvantages, the manual tuning of the admissible set, and the limited exploration.

Admissible Set. The worst-case optimization increases the policy robustness but also leads to very conservative policies. Especially when the admissible set is set too large, the policy becomes too pessimistic and does nothing. Furthermore, the admissible set is constant on the complete state domain. The constant magnitude of the disturbances can lead to state-space regions where the adversary is too powerful. For example, the swing-up of the physical Furuta pendulum is much more reliable with the adversary than without. However, balancing the pendulum using the same adversary yields a too conservative policy. On the Furuta pendulum, the robust policy caused chattering during the balancing. Therefore, it would be beneficial to have adaptive admissible sets in different regions of the state space. To avoid the manual tuning of the admissible set and prevent overly conservative policies, the admissible set could be learned from data of the physical system. This approach would very similar to existing domain randomization approaches, which learn the distributions of the randomized parameters from data. For domain randomization, this identification of the distributions has improved the performance [5, 20–22].

Exploration. rFVI performed well when used for dynamic programming on the complete state domain. However, dynamic programming is a strong limitation as it prevents scaling rFVI to more high-dimensional tasks. To scale the proposed approach to a higher dimensional system one would need to improve the exploration to enable real-time dynamic programming. Naive random exploration is too pessimistic for rFVI and does not find the solution as the magnitude of the admissible set is increased. Furthermore, exploration is hard in this setting as the optimal policy approximates a continuous-time policy. Hence, one wants to control the system with a high sampling frequency. The disadvantage of this high control frequency is that the frequent updates average out of the random exploration. Therefore, random exploration does not explore the state space. To improve the exploration one could combine the presented approach with online planning that is optimistically biased and ignores the adversary. In this case, one would explore the important regions of the state space and prevent the collapse of the state distribution. Furthermore, one could use the epistemic uncertainty of the value function ensembles to determine under-explored regions and add reward bonuses to uncertain regions.

References

1. Allen-Blanchette, C., Veer, S., Majumdar, A., Leonard, N.E.: Lagnetvip: a Lagrangian neural network for video prediction (2020). arXiv:2010.12932
2. Atkeson, C.G., An, C.H., Hollerbach, J.M.: Estimation of inertial parameters of manipulator loads and links. Int. J. Robot. Res. (1986)
3. de Avila Belbute-Peres, F., Smith, K., Allen, K., Tenenbaum, J., Kolter, J.Z.: End-to-end differentiable physics for learning and control. In: Advances in Neural Information Processing Systems (2018)
4. Botev, A., Jaegle, A., Wirnsberger, P., Hennes, D., Higgins, I.: Which priors matter? Benchmarking models for learning latent dynamics (2021). arXiv:2111.05458
5. Chebotar, Y., Handa, A., Makoviychuk, V., Macklin, M., Issac, J., Ratliff, N., Fox, D.: Closing the sim-to-real loop: adapting simulation randomization with real world experience. In: International Conference on Robotics and Automation (ICRA) (2019)
6. Cranmer, M., Greydanus, S., Hoyer, S., Battaglia, P., Spergel, D., Ho, S.: Lagrangian neural networks (2020). arXiv:2003.04630
7. Cranmer, M., Sanchez-Gonzalez, A., Battaglia, P., Xu, R., Cranmer, K., Spergel, D., Ho, S.: Discovering symbolic models from deep learning with inductive biases (2020). arXiv:2006.11287
8. Degrave, J., Hermans, M., Dambre, J., et al.: A differentiable physics engine for deep learning in robotics. Front. Neurorobot. (2019)
9. Gautier, M.: Identification of robots dynamics. IFAC Proc. Vol. (1986)
10. Geilinger, M., Hahn, D., Zehnder, J., Bächer, M., Thomaszewski, B., Coros, S.: Add: analytically differentiable dynamics for multi-body systems with frictional contact (2020). arXiv:2007.00987
11. Greydanus, S., Dzamba, M., Yosinski, J.: Hamiltonian neural networks. In: Advances in Neural Information Processing Systems (NeuRIPS) (2019)
12. Gupta, J.K., Menda, K., Manchester, Z., Kochenderfer, M.J.: A general framework for structured learning of mechanical systems (2019). arXiv:1902.08705
13. Heiden, E., Millard, D., Coumans, E., Sukhatme, G.S.: Augmenting differentiable simulators with neural networks to close the sim2real gap (2020). arXiv:2007.06045
14. Hochlehnert, A., Terenin, A., Sæmundsson, S., Deisenroth, M.: Learning contact dynamics using physically structured neural networks. In: International Conference on Artificial Intelligence and Statistics (Aistats) (2021)
15. Hu, Y., Anderson, L., Li, T.M., Sun, Q., Carr, N., Ragan-Kelley, J., Durand, F.: Difftaichi: Differentiable programming for physical simulation (2019). arXiv:1910.00935
16. Khosla, P.K., Kanade, T.: Parameter identification of robot dynamics. In: Conference on Decision and Control (CDC) (1985)
17. Lambert, N., Amos, B., Yadan, O., Calandra, R.: Objective mismatch in model-based reinforcement learning (2020). arXiv:2002.04523
18. Lutter, M., Hasenclever, L., Byravan, A., Dulac-Arnold, G., Trochim, P., Heess, N., Merel, J., Tassa, Y.: Learning dynamics models for model predictive agents (2021). arXiv:2109.14311
19. Mukerjee, A., Ballard, D.: Self-calibration in robot manipulators. In: International Conference on Robotics and Automation (ICRA) (1985)
20. Muratore, F., Eilers, C., Gienger, M., Peters, J.: Data-efficient domain randomization with bayesian optimization. IEEE Robot. Autom. Lett. (RAL) (2021)
21. Muratore, F., Gruner, T., Wiese, F., Belousov, B., Gienger, M., Peters, J.: Neural posterior domain randomization. In: Conference on Robot Learning (CoRL) (2021)
22. Ramos, F., Possas, R.C., Fox, D.: Bayessim: adaptive domain randomization via probabilistic inference for robotics simulators (2019). arXiv:1906.01728
23. Saemundsson, S., Terenin, A., Hofmann, K., Deisenroth, M.: Variational integrator networks for physically structured embeddings. In: International Conference on Artificial Intelligence and Statistics (Aistats) (2020)
24. Toth, P., Rezende, D.J., Jaegle, A., Racanière, S., Botev, A., Higgins, I.: Hamiltonian generative networks. International Conference on Learning Representations (ICLR) (2019)

25. Werling, K., Omens, D., Lee, J., Exarchos, I., Liu, C.K.: Fast and feature-complete differentiable physics for articulated rigid bodies with contact (2021). arXiv:2103.16021
26. Zhong, Y.D., Dey, B., Chakraborty, A.: Symplectic ode-net: learning hamiltonian dynamics with control (2019). arXiv:1909.12077
27. Zhong, Y.D., Dey, B., Chakraborty, A.: A differentiable contact model to extend Lagrangian and Hamiltonian neural networks for modeling hybrid dynamics (2021). arXiv:2102.06794